# PE Environmental
## Practice Exams

R. Wane Schneiter, PhD, PE

Professional Publications, Inc. • Belmont, California

## Benefit by Registering This Book with PPI

- Get book updates and corrections.
- Hear the latest exam news.
- Obtain exclusive exam tips and strategies.
- Receive special discounts.

Register your book at **ppi2pass.com/register**.

## Report Errors and View Corrections for This Book

PPI is grateful to every reader who notifies us of a possible error. Your feedback allows us to improve the quality and accuracy of our products. You can report errata and view corrections at **ppi2pass.com/errata**.

**PE ENVIRONMENTAL PRACTICE EXAMS**

Current release of this edition: 1

**Release History**

| date | edition number | revision number | update |
|---|---|---|---|
| Apr 2018 | 1 | 1 | New edition. |

© 2018 Professional Publications, Inc. All rights reserved.

All content is copyrighted by Professional Publications, Inc. (PPI). No part, either text or image, may be used for any purpose other than personal use. Reproduction, modification, storage in a retrieval system or retransmission, in any form or by any means, electronic, mechanical, or otherwise, for reasons other than personal use, without prior written permission from the publisher is strictly prohibited. For written permission, contact PPI at permissions@ppi2pass.com.

Printed in the United States of America.

PPI
1250 Fifth Avenue, Belmont, CA 94002
(650) 593-9119
ppi2pass.com

ISBN: 978-1-59126-574-0

Library of Congress Control Number: 201839439

F E D C B A

# Table of Contents

**PREFACE** .................................................................................................................. v

**INTRODUCTION** ...................................................................................................... vii

**PRACTICE EXAM 1** ................................................................................................... 1
    Morning Session Answer Sheet 1 ........................................................................ 1
    Morning Session 1 .................................................................................................. 3
    Afternoon Session Answer Sheet 1 ...................................................................... 11
    Afternoon Session 1 .............................................................................................. 13

**PRACTICE EXAM 2** ................................................................................................. 21
    Morning Session Answer Sheet 2 ........................................................................ 21
    Morning Session 2 ................................................................................................ 23
    Afternoon Session Answer Sheet 2 ...................................................................... 33
    Afternoon Session 2 .............................................................................................. 35

**ANSWER KEY** ......................................................................................................... 45
    Exam 1 Answer Key ............................................................................................. 45
    Exam 2 Answer Key ............................................................................................. 47

**SOLUTIONS** ............................................................................................................ 49
    Morning Session 1 ................................................................................................ 49
    Afternoon Session 1 .............................................................................................. 57
    Morning Session 2 ................................................................................................ 67
    Afternoon Session 2 .............................................................................................. 75

# Preface

The National Council of Examiners for Engineering and Surveying (NCEES) prepares the Principles and Practice of Engineering (PE) examination for environmental engineering from problems submitted by professional engineers representing consulting, government, and industry. The exam provides a uniform tool for local licensing boards to assess the competency of engineers practicing within their jurisdictions. NCEES does not provide copies of past exams, and it keeps exam problems confidential so that actual problems included on any NCEES exam cannot be included in *PE Environmental Practice Exams* or any other study guide. However, NCEES does identify the general subject areas covered on the exam.

Within the limitations imposed by NCEES, *PE Environmental Practice Exams* presents a broad range of problems relevant to those that may be encountered in the practice of environmental engineering. Organized as two separate practice exams of 80 problems each, *PE Environmental Practice Exams* addresses those general subject areas identified by NCEES and covers a broad range of topics, both conceptual and practical, with varying levels of difficulty, and in a variety of forms. Although you will not encounter any problems on the exam exactly like those presented in this book, the problems presented here are believed to be representative of the type and difficulty of those you will encounter on the exam. An eight-hour exam of 80 problems allows six minutes to complete each problem. Some problems may require only several seconds to complete while others may take considerably more than six minutes. You will find this to be true of the problems in this book. The problems requiring more time will, in particular, enhance your exam preparation.

*PE Environmental Practice Exams* should be used to practice solving engineering problems in a test format and to evaluate your overall preparedness for taking the exam. You should not rely on *PE Environmental Practice Exams* as your only study guide to prepare for the exam. You should instead combine its use with the use of other study references. Beginning with April 2019 exam, the only resource material available to you during the exam will be the most current version of the *NCEES PE Environmental Reference Handbook*. Familiarity with the *Handbook* will be important to your success. Use it when solving practice problems.

Review environmental engineering principles and concepts before attempting the sample problems, and then use these sample problems to determine your weaknesses and to direct subsequent study. This will enable you to focus your efforts where they will be most productive, to practice solving exam-like problems.

Solutions presented for each practice problem may represent only one of several alternative methods for obtaining a correct answer. It may also be possible that an alternative method of solving a problem will produce a different, but still appropriate, answer.

Although great care was exercised in preparing *PE Environmental Practice Exams* to ensure that representative problems were included and that they were solved correctly, some errors may exist. I ask for your assistance in improving subsequent editions by bringing any errors to my attention through PPI's website at **ppi2pass.com/errata**, and by providing comments regarding the applicability of specific problems to those you encounter on the exam.

Good luck!

R. Wane Schneiter, PhD, PE

# Introduction

## EXAM ORGANIZATION

The PE Environmental exam consists of two parts, a morning and an afternoon session, each lasting four hours. In completing the exam you will need to provide answers to 80 multiple-choice problems, 40 in the morning and 40 in the afternoon.

The exam uses both the International System (SI) and the U.S. Customary System of units.

Four possible answers are provided for each problem, but only one of the possible answers is correct. One point is awarded for each correct answer. No partial credit is given, but no penalty is assessed for incorrect answers. However, if more than one choice is marked for a single problem, no points will be awarded. The minimum passing score is not released by the NCEES, as it differs for each exam administration.

The morning and afternoon sessions will include problems that address topics in the following categories, with the approximate number of questions indicated.

| | |
|---|---|
| **Water** | **24 questions** |
| A. Principles | 4 questions |
| B. Wastewater | 8 questions |
| C. Stormwater | 2 questions |
| D. Potable Water | 8 questions |
| E. Water Resources | 2 questions |
| **Air** | **16 questions** |
| A. Principles | 8 questions |
| B. Pollution Control | 8 questions |
| **Solid and Hazardous Waste** | **12 questions** |
| A. Principles | 6 problems |
| B. Municipal and Industrial Solid Waste | 4 problems |
| C. Hazardous, Medical, and Radioactive Waste | 2 problems |
| **Site Assessment and Remediation** | **14 problems** |
| A. Principles | 6 problems |
| B. Applications | 8 problems |
| **Environmental Health and Safety** | **8 problems** |
| A. Principles | 4 problems |
| B. Applications | 4 problems |
| **Associated Engineering Principles** | **6 problems** |
| A. Principles | 2 problems |
| B. Applications | 4 problems |
| **Total** | **80 problems** |

For the October 2018 exam, both the morning and the afternoon sessions of the exam are open book. In general, any bound reference material is allowed, including personal notes and sample calculations. Textbooks, handbooks, and other professional reference books are allowed. The *NCEES PE Environmental Reference Handbook* may be used on exam day, but you will need to bring your personal bound paper copy. All solutions to the exam problems must be marked on the multiple-choice answer sheet provided. Any notes or calculations marked in the exam booklet will not be considered as part of a solution and will not be graded. The multiple-choice answer sheet is machine-graded and is the only record that will be scored.

Beginning with the April 2019 exam, the exam will be a computer based test (CBT) and the most current version of the *NCEES PE Environmental Reference Handbook* will be the only reference material allowed for use during the exam. The handbook will be available for your use in a searchable PDF computer-based format.

Additionally, in the PE Environmental CBT exam, no personal writing tablets, scratch paper, or other unbound notes or materials will be permitted. Mechanical pencils and scratch paper will be provided at the test site. Battery-operated, silent, nonprinting calculators are allowed. Calculators with communication or text editing capabilities are not allowed. Each local jurisdiction defines specifically which materials you are permitted to bring with you to the exam, so check with them early to ensure that you are compiling acceptable materials for use during the exam. Information about the state boards can be found at **ppi2pass.com/stateboards**.

The exam is intended to assess your personal competence without discussing or sharing information with others during the exam period. Do not expect to communicate with others while taking the exam.

For the latest environmental engineering exam resources, check the PPI website at **ppi2pass.com/shop/pe-exam/environmental-pe-exam**.

# Morning Session Answer Sheet 1

| | | | | |
|---|---|---|---|---|
| 1. (A) (B) (C) (D) | 11. (A) (B) (C) (D) | 21. (A) (B) (C) (D) | 31. (A) (B) (C) (D) |
| 2. (A) (B) (C) (D) | 12. (A) (B) (C) (D) | 22. (A) (B) (C) (D) | 32. (A) (B) (C) (D) |
| 3. (A) (B) (C) (D) | 13. (A) (B) (C) (D) | 23. (A) (B) (C) (D) | 33. (A) (B) (C) (D) |
| 4. (A) (B) (C) (D) | 14. (A) (B) (C) (D) | 24. (A) (B) (C) (D) | 34. (A) (B) (C) (D) |
| 5. (A) (B) (C) (D) | 15. (A) (B) (C) (D) | 25. (A) (B) (C) (D) | 35. (A) (B) (C) (D) |
| 6. (A) (B) (C) (D) | 16. (A) (B) (C) (D) | 26. (A) (B) (C) (D) | 36. (A) (B) (C) (D) |
| 7. (A) (B) (C) (D) | 17. (A) (B) (C) (D) | 27. (A) (B) (C) (D) | 37. (A) (B) (C) (D) |
| 8. (A) (B) (C) (D) | 18. (A) (B) (C) (D) | 28. (A) (B) (C) (D) | 38. (A) (B) (C) (D) |
| 9. (A) (B) (C) (D) | 19. (A) (B) (C) (D) | 29. (A) (B) (C) (D) | 39. (A) (B) (C) (D) |
| 10. (A) (B) (C) (D) | 20. (A) (B) (C) (D) | 30. (A) (B) (C) (D) | 40. (A) (B) (C) (D) |

# Morning Session 1

**1.** The average annual wastewater flow for a combined commercial/residential development is $9 \times 10^6$ gal/yr. The typical daily flow during the maximum month is 125% of the average daily flow during the year. What is most nearly the daily average wastewater flow?

(A) 31,000 gal/day
(B) 35,000 gal/day
(C) 45,000 gal/day
(D) 68,000 gal/day

**2.** What is most nearly the required storage volume for a runoff volume of 1.2 cm and a storage-to-runoff-volume ratio of 0.25?

(A) 0.25 cm
(B) 0.30 cm
(C) 1.2 cm
(D) 3.3 cm

NOT IN HAND BOOK

**3.** Four parallel sedimentation basins are required to treat 12 000 m³/d. What is most nearly the overflow rate for a settling zone depth of 2.5 m and a settling time of 100 min?

(A) 0.38 m³/m²·h
(B) 0.75 m³/m²·h
(C) 0.96 m³/m²·h
(D) 1.5 m³/m²·h

**4.** Two sedimentation basins have a total flow rate of 7500 m³/d and a weir overflow rate of 14 m³/m·h. What is most nearly the weir length required for each of the two sedimentation basins?

(A) 4.5 m/tank
(B) 11 m/tank
(C) 23 m/tank
(D) 140 m/tank

**5.** The influent to a wastewater treatment pond contains 200 mg/L biochemical oxygen demand (BOD), 15 mg/L total Kjeldahl nitrogen (TKN) and 4 mg/L total phosphorus (TP). Carbon is represented by BOD, and the typical C:N:P ratio for bacterial cells is 60:12:1. What measures are justified for the pond?

(A) augmentation of nitrogen
(B) augmentation of phosphorous
(C) provisions for removal of nitrogen and phosphorous
(D) no special measures for removal or augmentation are justified

**6.** A river receiving a wastewater discharge has the following characteristics.

| | |
|---|---|
| temperature | 48°F |
| dissolved oxygen (DO) upstream of discharge point | 8.9 mg/L |
| deoxygenation constant | 0.16/d |
| reoxygenation constant | 0.23/d |

What is most nearly the critical time for a mixed flow ultimate biochemical oxygen demand (BOD$_u$) at the discharge point of 10 mg/L?

(A) 1.7 d
(B) 2.6 d
(C) 3.4 d
(D) 6.4 d

**7.** A public water supply is disinfected with chlorine to inactivate giardia cysts. The chlorine feed occurs as the water enters the wet well where it experiences a minimum hydraulic residence time of 30 min. The free chlorine concentration is 2.0 mg/L. The water is at 15°C and 16.5 pH. What is most nearly the minimum hydraulic residence time that is required to achieve a 3-log reduction in giardia cysts?

(A) 19 min
(B) 35 min
(C) 38 min
(D) 69 min

**8.** A complete mix activated sludge process has been selected for treatment of a wastewater. Available design information is presented in the following table.

| parameter | value |
|---|---|
| flow rate from primary clarifiers | 5000 m³/d |
| influent substrate | 192 mg/L $BOD_5$ |
| effluent substrate | 20 mg/L soluble $BOD_5$ |
| yield coefficient | 0.5 at 20°C |
| observed yield coefficient | 0.36 at 20°C |
| minimum solids residence time | 3.0 d |
| activated sludge safety factor | 2.5 |

What is most nearly the daily dry mass of biosolids produced in the bioreactor?

(A) 310 kg/d
(B) 320 kg/d
(C) 430 kg/d
(D) 450 kg/d

**9.** A welded steel pipeline 1373 ft long is needed to convey 0.5 ft³/sec of water with a maximum allowable head loss of 1.2 ft. Most nearly, what standard pipe diameter is required?

(A) 6 in
(B) 8 in
(C) 10 in
(D) 20 in

**10.** Results of dye tracer studies conducted to define the flow characteristics of a reaction tank for a water process are tabulated and graphed in the following table and figure. The reactor volume was 25,000 gal, and the flow rate to the reactor was 500 gal/min.

| time (min) | concentration ($\mu$g/L) |
|---|---|
| 20 | 0 |
| 30 | 100 |
| 40 | 390 |
| 50 | 148 |
| 60 | 83 |
| 70 | 47 |
| 80 | 25 |
| 90 | 12 |
| 100 | 7.5 |
| 110 | 2.5 |

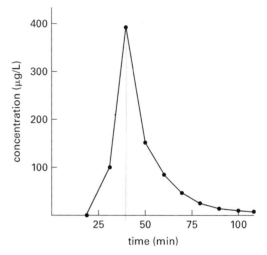

What is most nearly the reactor hydraulic residence time that corresponds to the peak of the concentration time plot?

(A) 10 min
(B) 26 min
(C) 40 min
(D) 47 min

**11.** A city recreation facility is being proposed that would occupy 20 ac. Twenty percent of the site will be landscaped and will require irrigation 26 wk/yr with a typical lawn sprinkler rate of 1 in/wk. What is most nearly the average annual irrigation water demand?

(A) 2,800,000 gal/yr
(B) 5,700,000 gal/yr
(C) 11,000,000 gal/yr
(D) 14,000,000 gal/yr

**12.** A residential/commercial development is proposed that would consist of the following facilities.

- 100-unit studio and single bedroom residential low-rise apartments with anticipated average occupancy of two persons per apartment and anticipated water use of 100 gal/person-day
- building with commercial space to accommodate 100 office employees, with an anticipated water use of 15 gal/person-day
- a full-service restaurant that will serve 62 lunch and 62 dinner customers daily and operate 7 day/wk, with an anticipated water use of 9 gal/person-day
- a luncheon deli that will serve 36 customers daily during the 5-day business week, with an anticipated water use of 6 gal/person-day
- a tennis and swimming club for office employees and apartment residents (40% expected use rate), with an anticipated water use of 100 gal/person-day

What is most nearly the average annual water demand for the development?

(A) 9,000,000 gal/yr
(B) 12,000,000 gal/yr
(C) 19,000,000 gal/yr
(D) 46,000,000 gal/yr

**13.** A horizontal paddle-wheel type flocculator is needed to treat a design flow of 5000 m³/d. The average velocity gradient over two sections is 45 s⁻¹. What is most nearly the total power required at the flocculator paddles, assuming a single, two-section, 100 m³ tank is used?

(A) 0.049 kW
(B) 0.097 kW
(C) 0.20 kW
(D) 0.24 kW

**14.** Results of soluble biochemical oxygen demand (BOD) analyses of stream water are as follows. Standardized testing procedures were used.

| sample | volume (mL) | $DO_1$ (mg/L) | $DO_5$ (mg/L) |
|---|---|---|---|
| 1 | 200 | 9.1 | 1.6 |
| 2 | 100 | 9.2 | 2.3 |
| 3 | 50 | 9.3 | 5.8 |
| 4 | 20 | 9.3 | 7.2 |

The deoxygenation rate constant at 25°C, $K_d$ (base $e$), is 0.40/d, and the temperature correction coefficient, $\theta_c$, is 1.047. All samples were incubated at 25°C for 5 d.

What is most nearly the 5-day biochemical oxygen demand ($BOD_5$) concentration at 20°C for the wastewater?

(A) 17 mg/L
(B) 19 mg/L
(C) 21 mg/L
(D) 24 mg/L

**15.** A wastewater discharge to a river produces the following dissolved oxygen (DO) sag curve. The temperature of the river water is 50°F.

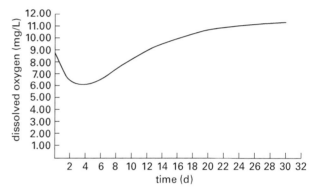

What is most nearly the critical DO deficit?

(A) 2.2 mg/L
(B) 3.9 mg/L
(C) 5.2 mg/L
(D) 6.1 mg/L

**16.** Two adsorber vessels, each with 20,000 lbm granular activated carbon (GAC) capacity, are used in a lead-follow configuration to treat contaminated groundwater. What is most nearly the GAC change-out interval per adsorption vessel for a GAC use rate of 500 lbm/day?

(A) 15 days/vessel
(B) 20 days/vessel
(C) 40 days/vessel
(D) 80 days/vessel

**17.** Does a developed property likely require stormwater control and containment devices if the pre-development flow is 1 m³/s and the post-development flow is 1.5 m³/s?

(A) yes, because control and containment are always provided when post-development exceeds pre-development flow

(B) yes, because post-development flow is considerably greater than pre-development flow

(C) no, because post-development flow is less than twice pre-development flow

(D) no, because post-development and pre-development flows are very small

**18.** An industrial discharger is unable to satisfy the permit pre-treatment requirements for discharge to the local municipal wastewater treatment plant. The current wastewater five-day biochemical oxygen demand ($BOD_5$) concentration of 2000 mg/L for a wastewater flow is 400 m³/d. The pre-treatment system consists of three ponds operating in series. Pond 1 is an anaerobic pond and ponds 2 and 3 are intended to operate as aerobic ponds. A schematic of the pond system follows.

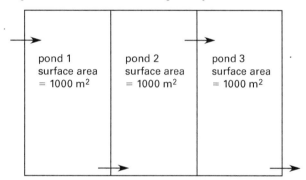

Assuming a 50% BOD removal efficiency in pond 1 and a 1-to-1 ratio of dissolved oxygen required for BOD removal, what would most nearly be the minimum theoretical aeration requirement for an additional 80% BOD removal through pond 2?

(A) 13 kg DO/h
(B) 16 kg DO/h
(C) 27 kg DO/h
(D) 33 kg DO/h

**19.** A sedimentation basin is required to treat 10 000 m³/d of flow containing 234 mg/L TSS. The settling zone depth is 2.5 m, the settling time is 100 min, and the settling zone length-to-width ratio is 3:1. Settling characteristics are as described in the following illustration.

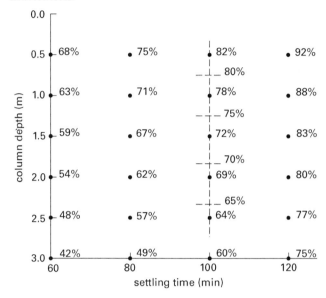

What is the effect on efficiency if the settling zone depth is decreased to 2.0 m?

(A) decreases
(B) increases
(C) remains unchanged if settling time remains unchanged
(D) remains unchanged if overflow rate remains unchanged

**20.** A 200 mL water sample was titrated with 0.03N $H_2SO_4$. The initial pH was 7.8, and 28 mL of acid was added to reach pH 4.5. What are the dominant alkalinity species present in the sample?

(A) carbonate only
(B) bicarbonate only
(C) bicarbonate and carbonate
(D) bicarbonate, carbonate, and hydroxide

**21.** Water samples were collected and submitted to a commercial analytical laboratory for analysis. The laboratory provided the following results.

| ion | concentration (mg/L) |
|---|---|
| $SO_4^{-2}$ | 17 |
| $Na^+$ | 48 |
| $K^+$ | 14 |
| $Cl^-$ | 39 |
| $Ca^{+2}$ | 56 |
| $Mg^{+2}$ | 12 |
| $HCO_3^-$ | 237 |

Is the ion analysis complete?

(A) yes

(B) no, it is deficient in anions

(C) no, it is deficient in cations

(D) cannot be determined

**22.** A water sample has a total dissolved solids (TDS) concentration of 425 mg/L and a specific conductivity of 32 $\mu$S. What materials likely comprise the majority of the TDS of the sample?

(A) nonionic solids

(B) ionic solids

(C) approximately equal amounts of both ionic and nonionic solids

(D) cannot be determined

**23.** A small community plans to discharge 0.25 MGD of wastewater for treatment using a single-stage trickling filter. The wastewater has a $BOD_5$ concentration of 186 mg/L after primary treatment. A $BOD_5$ concentration of 30 mg/L is required at the filter discharge. For a recirculation factor of 2, what is most nearly the total volume of filter media required?

(A) 2300 $ft^3$

(B) 14,000 $ft^3$

(C) 70,000 $ft^3$

(D) 125,000 $ft^3$

**24.** A wastewater treatment plant treats 2.5 millions of gallons per day (MGD) of wastewater containing 42 mg/L nitrate nitrogen and 13 mg/L nitrite nitrogen.

To further reduce the nitrogen concentrations through denitrification, the plant will use methanol in a tertiary treatment process. The specific gravity of methanol is 0.7915. Most nearly, what is the daily volume of methanol required if the initial dissolved oxygen concentration is 1.9 mg/L?

(A) 110 L/d

(B) 180 L/d

(C) 930 L/d

(D) 1490 L/d

**25.** The normal chemical composition of a dry gas is given in the following table.

| component | concentration (%) |
|---|---|
| nitrogen | 53 |
| oxygen | 31 |
| methane | 12 |
| carbon dioxide | 4 |

The atmospheric pressure is 0.98 atm. What is most nearly the partial pressure exerted by the methane?

(A) 0.0817 atm

(B) 0.118 atm

(C) 0.770 atm

(D) 0.860 atm

**26.** What is the difference between primary and secondary standards, as used with the National Ambient Air Quality Standards (NAAQS), and primary and secondary air pollutants?

(A) Primary and secondary standards define the permissible levels of primary and secondary pollutants, respectively, that can exist in ambient air.

(B) Primary and secondary standards are regulatory levels to protect human health and prevent environmental damage, respectively. Primary and secondary pollutants relate to the emitted form of the pollutants.

(C) Primary and secondary standards refer to the level of treatment required by air pollution control equipment. Primary and secondary pollutants define the type of emissions from primary and secondary treatment equipment.

(D) There is no difference. They refer to the same thing.

**27.** An electrostatic precipitator (ESP) is being considered for use as an air pollution control process for a metal parts manufacturer. The air and particle characteristics are defined by the following parameters.

| air flow rate requiring treatment | 14 m³/s |
| particulate concentration | 20 g/m³ |

What is most nearly the daily mass of particulate removed by the ESP if operated at 80% efficiency?

(A) 3500 kg/d
(B) 7000 kg/d
(C) 14 000 kg/d
(D) 19 000 kg/d

**28.** Most nearly, what is the dry air density for a barometric pressure of 29.81 in Hg at 58°F?

(A) 0.056 lbf/ft³
(B) 0.067 lbf/ft³
(C) 0.076 lbf/ft³
(D) 0.086 lbf/ft³

**29.** A power plant uses coal with a sulfur content of 2.7% at a feed rate of 8.3 lbm/sec. What is most nearly the daily sulfur (as $SO_2$) emitted to air pollution control (APC) equipment if 5% of the sulfur bypasses the APC?

(A) 1900 lbm/day
(B) 9200 lbm/day
(C) 18,000 lbm/day
(D) 37,000 lbm/day

**30.** What emitted sulfur and nitrogen compounds most often occur as precursors to secondary pollutants?

(A) nitric oxide and elemental sulfur
(B) nitric oxide and sulfur dioxide
(C) nitrogen dioxide and sulfur dioxide
(D) nitrogen dioxide and sulfur trioxide

**31.** A cross current-flow scrubber used for air pollution control has a spray nozzle pressure of 0.05 atm. What is most nearly the air pressure drop through the scrubber spray nozzle?

(A) 0.005 atm
(B) 0.05 atm
(C) 0.5 atm
(D) 1.0 atm

**32.** The following illustration represents atmospheric stability conditions at a site with a ground surface elevation of 100 m.

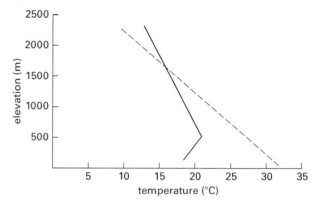

Which phrase best describes the atmospheric conditions?

(A) inversion to 500 m, subadiabatic above 500 m
(B) inversion to 500 m, superadiabatic above 500 m
(C) subadiabatic to 500 m, superadiabatic above 500 m
(D) superadiabatic to 500 m, subadiabatic above 500 m

**33.** A stack with an effective height of 150 m emits pollutants at 5.4 kg/s. Wind speed measured 10 m above the ground surface is 2.6 m/s on a sunny day with moderate incoming solar radiation. For pollutants emitted from the stack, what is most nearly the maximum ground-level concentration downwind of the source?

(A) 4 mg/m³
(B) 12 mg/m³
(C) 17 mg/m³
(D) 42 mg/m³

**34.** What is the bubble policy for using emission reduction credits (ERC) under the emission trading program?

(A) allowing industrial expansion in non-attainment areas by letting industry compensate for new emissions by offsetting them with ERC acquired from other industries existing in the region

(B) treating all activities in a single plant or among proximate industries as a group, and allowing emissions at varied rates from the group's sources as long as the total emissions do not exceed those allowed for each separate source

(C) saving ERC for future use

(D) allowing industry to expand without acquiring a new permit as long as the net increase in emissions is not significant

**35.** A baghouse is used to remove particulate from a waste air stream. The baghouse and particulate characteristics are

| superficial filtering velocity | 3.5 m/min |
| dust loading | 25 g/m$^3$ |
| particle layer bulk density | $1.2 \times 10^6$ g/m$^3$ |

What is most nearly the depth of the dust layer after 1 h of operation?

(A) 0.44 cm
(B) 0.72 cm
(C) 1.5 cm
(D) 2.3 cm

**36.** A pulsed-air baghouse is used to control particulate from a manufacturing facility. What is most nearly the total fabric area required for the baghouse with eight compartments if it has eight compartments and each compartment has 60 bags of 10 m$^2$ surface area each?

(A) 130 m$^2$
(B) 750 m$^2$
(C) 4800 m$^2$
(D) 6000 m$^2$

**37.** Which of the following is generally NOT characteristic of electrostatic precipitators?

(A) economical
(B) efficient for moist flows and mists
(C) reliable and predictable
(D) removal efficiencies near 99%

**38.** A plume at 31°C is emitted from a 93 m high stack where the ground-level air temperature is 19°C. The ambient lapse rate is −0.0047°C/m. Most nearly how high will the plume rise?

(A) 2170 m
(B) 2250 m
(C) 2430 m
(D) 2520 m

**39.** What is radon gas?

(A) a naturally occurring radioactive gas that can cause lung cancer
(B) a volatile organic compound that can cause adverse health effects
(C) a synthetic radioactive compound emitted to the atmosphere from nuclear reactor cooling water
(D) a radioactive gas characterized by pungent taste and odor

**40.** An above-ground storage tank containing a toxic volatile chemical is ruptured, and 10 kg of vapor is emitted to the surrounding air. The release occurs at night with a clear sky and 2.1 m/s wind speed. The nearest residence is located 1.4 km directly downwind. The terrain is flat, with no significant obstacles present between the release area and the residence. Approximately how much time is available for the residents to evacuate their homes to avoid exposure to the vapor?

(A) < 2 min
(B) < 11 min
(C) < 25 min
(D) < 49 min

# STOP!

### DO NOT CONTINUE!

This concludes the Morning Session of the examination. If you finish early, check your work and make sure that you have followed all instructions. After checking your answers, submit your solutions and leave the examination room. Once your answers are submitted you will not be able to access them again.

# Afternoon Session Answer Sheet 1

41. (A) (B) (C) (D)    51. (A) (B) (C) (D)    61. (A) (B) (C) (D)    71. (A) (B) (C) (D)
42. (A) (B) (C) (D)    52. (A) (B) (C) (D)    62. (A) (B) (C) (D)    72. (A) (B) (C) (D)
43. (A) (B) (C) (D)    53. (A) (B) (C) (D)    63. (A) (B) (C) (D)    73. (A) (B) (C) (D)
44. (A) (B) (C) (D)    54. (A) (B) (C) (D)    64. (A) (B) (C) (D)    74. (A) (B) (C) (D)
45. (A) (B) (C) (D)    55. (A) (B) (C) (D)    65. (A) (B) (C) (D)    75. (A) (B) (C) (D)
46. (A) (B) (C) (D)    56. (A) (B) (C) (D)    66. (A) (B) (C) (D)    76. (A) (B) (C) (D)
47. (A) (B) (C) (D)    57. (A) (B) (C) (D)    67. (A) (B) (C) (D)    77. (A) (B) (C) (D)
48. (A) (B) (C) (D)    58. (A) (B) (C) (D)    68. (A) (B) (C) (D)    78. (A) (B) (C) (D)
49. (A) (B) (C) (D)    59. (A) (B) (C) (D)    69. (A) (B) (C) (D)    79. (A) (B) (C) (D)
50. (A) (B) (C) (D)    60. (A) (B) (C) (D)    70. (A) (B) (C) (D)    80. (A) (B) (C) (D)

# Afternoon Session 1

**41.** The combustion reaction for methane is

$$CH_4 + 2O_2 \rightarrow CO_2 + 2H_2O$$

The standard enthalpies are

$\Delta H°\ CH_4$      $-74.82$ kJ/mol

$\Delta H°\ O_2$      0 (standard state)

$\Delta H°\ CO_2$      $-393.5$ kJ/mol

$\Delta H°\ H_2O$      $-241.8$ kJ/mol

What is most nearly the heating value of the methane per 1000 ft³ of methane at 20°C and 1 atm?

(A) $8.6 \times 10^1$ kJ/1000 ft³

(B) $9.3 \times 10^2$ kJ/1000 ft³

(C) $8.6 \times 10^4$ kJ/1000 ft³

(D) $1.0 \times 10^6$ kJ/1000 ft³

**42.** Which incinerator type presents the most difficulty in meeting air emission limits?

(A) controlled air

(B) rotary kiln

(C) fluid bed

(D) multiple hearth

**43.** Solid waste generated by a municipality is characterized as follows.

| waste component | mass (%) | component discarded energy (kJ/kg) | ash (%) |
|---|---|---|---|
| food | 13 | 4650 | 5 |
| glass | 6 | 150 | 98 |
| plastic | 4 | 32,600 | 10 |
| paper | 37 | 16,750 | 6 |
| cardboard | 10 | 16,300 | 5 |
| textiles | 1 | 17,450 | 2.5 |
| ferrous metal | 8 | 700 | 98 |
| non-ferrous metal | 2 | 700 | 96 |
| wood | 4 | 18,600 | 1.5 |
| yard clippings | 15 | 6500 | 4.5 |

What is most nearly the energy content of the discarded bulk waste without ash?

(A) $1.5 \times 10^7$ kJ/1000 kg

(B) $3.6 \times 10^7$ kJ/1000 kg

(C) $4.3 \times 10^7$ kJ/1000 kg

(D) $4.9 \times 10^7$ kJ/1000 kg

**44.** A municipal solid waste is put in a landfill. The municipality producing the waste has 13 000 residents, who each generate the waste at approximately 1.2 kg/person·d. What is most nearly the landfill volume required for a 25 yr active life with a cover-to-waste ratio of 1:5 and a maximum compacted density of 1100 kg/m³?

(A) $160 \times 10^3$ m³

(B) $240 \times 10^3$ m³

(C) $1300 \times 10^3$ m³

(D) $2200 \times 10^3$ m³

**45.** An incinerator is to have a bed area loading of $2.3 \times 10^6$ kJ/m² h for a waste stream of 500 kg/h with a heating value of 57 000 kJ/kg. What is most nearly the minimum required incinerator bed area?

(A) 0.5 m²
(B) 4.3 m²
(C) 12 m²
(D) 26 m²

**46.** An activated sludge plant produces 4 m³/d of wasted sludge dewatered to 76% moisture that requires disposal. The specific heat of water is 4.184 kJ/kg·°C and the heat of vaporization is 2258 kJ/kg. Most nearly how much heat is required to dry the sludge at 100°C from 76% moisture to 0% moisture if the initial temperature of the sludge is 20°C?

(A) 43 000 kJ/h
(B) 240 000 kJ/h
(C) 290 000 kJ/h
(D) 330 000 kJ/h

**47.** A groundwater source used in an industrial process has the following characteristics.

| chemical | concentration (mg/L) |
|---|---|
| $Ca^{+2}$ | 187 |
| $Mg^{+2}$ | 42 |
| $Na^+$ | 115 |
| $HCO_3^-$ | 714 |

Water is pumped at a uniform, continuous rate of 0.7 m³/min during two 8 h shifts, 20 d/mo. The precipitation reaction for calcium hardness using sodium hydroxide (NaOH) is

$$Ca^{+2} + 2HCO_3^- + 2NaOH \rightarrow CaCO_3 + Na_2CO_3 + 2H_2O$$

What is most nearly the annual dry mass of sludge produced from calcium hardness precipitation?

(A) 16 000 kg/yr
(B) 75 000 kg/yr
(C) 92 000 kg/yr
(D) 170 000 kg/yr

**48.** Wastewater at 25°C contains cadmium at 121 mg/L in a solution at pH 4.1. The solubility product for cadmium is $5.27 \times 10^{-15}$ at 25°C. What is most nearly the cadmium concentration if the pH is raised to 8.6?

(A) 19 mg/L
(B) 27 mg/L
(C) 37 mg/L
(D) 52 mg/L

**49.** A community of 37,000 residents produces solid waste at 2.2 lbm/person-day. The trucks available to collect the waste have a 12 yd³ capacity and can compact the waste to 760 lbm/yd³. The collection crews work one 8 h shift each day, Monday through Friday. What is the number of trucks required to collect all the waste in the community once weekly if any truck can complete three loads per day?

(A) three trucks
(B) five trucks
(C) seven trucks
(D) nine trucks

**50.** A municipality operates an anaerobic digester. To increase the value of the digester gases and reduce carbon dioxide emissions, the city is considering separating carbon dioxide from other digester gases and then selling it. The digester produces about 60,000 ft³ of gas daily at 1 atm and 25°C; about 25% of the gas is carbon dioxide. The molar gas volume is 24.4 L/mol at 25°C. What is most nearly the daily mass of carbon dioxide generated by the digester?

(A) 760 kg/d
(B) 830 kg/d
(C) 3300 kg/d
(D) 4200 kg/d

**51.** A facility produces 250,000 gal/day of wastewater containing heavy metals, including chromium. Ion exchange is used to recover the chromium and remove other metals from the wastewater. The hydraulic loading rate (HLR) is 0.48 ft³/ft³-min. What measures could be implemented to increase HLR?

(A) add exchanger vessels in parallel
(B) add exchanger vessels in series
(C) recirculate flow
(D) none of the above

**52.** A landfill liner measuring 3 ft thick has a porosity of 0.15 and a hydraulic conductivity of 0.00023 ft/day. If the leachate depth above the liner will rise to 2.3 ft, most nearly how long will it take for the leachate to pass through the liner?

(A) 1 yr
(B) 3 yr
(C) 36 yr
(D) 1100 yr

**53.** An electroplater produces wastewater at a continuous rate of 1.8 m$^3$/min with Cr$^{+6}$ (measured as CrO$_3$) at a concentration of 534 mg/L. Sodium bisulfite (NaHSO$_3$) and sulfuric acid (H$_2$SO$_4$) are selected to reduce the chromium from the hexavalent to the trivalent form. The chemical equation for the reduction reaction is

$$4CrO_3 + 6NaHSO_3 + 3H_2SO_4$$
$$\rightarrow 3Na_2SO_4 + 2Cr_2(SO_4)_3 + 6H_2O$$

Sodium bisulfite is available at 92% purity for $120/1000 kg. What is most nearly the annual cost of sodium bisulfite required for reduction, assuming the facility operates 24 h/d, 365 d/yr?

(A) $69,000/yr
(B) $95,000/yr
(C) $100,000/yr
(D) $600,000/yr

**54.** What is the primary difference in the dose-response relationship when comparing carcinogens with noncarcinogens?

(A) The slope of the dose-response curve is steeper for carcinogens than for noncarcinogens.
(B) The response from exposure to carcinogens is constant regardless of dose.
(C) A threshold dose exists for noncarcinogens below which no response is expected; there is no threshold for carcinogens.
(D) There are no general differences between the two.

**55.** Lifting weight limit studies have produced the results shown.

| | |
|---|---|
| horizontal distance of the hand from the midpoint of the line joining the inner ankle bones to a point projected on the floor directly below the load center | 17 in |
| vertical distance of the hands from the floor | 36 in |
| vertical distance traveled by the hands between the origin and destination of the lift | 44 in |
| asymmetry angle | 71° |

The lifting frequency is 0.5/min for up to 8 hr/day. The container being lifted is optimally designed but does not include cut-out handles. Most nearly, what is the recommended weight limit?

(A) 13 lbf
(B) 15 lbf
(C) 24 lbf
(D) There is no safe weight limit.

**56.** What is the level of exposure to a toxic chemical that a population can be assumed to endure without appreciable risk resulting?

(A) lowest observed adverse effect level (LOAEL)
(B) lowest observed effect level (LOEL)
(C) no observed effect level (NOEL)
(D) reference dose (RfD)

**57.** The quality factor for alpha particles is 20. For an absorbed alpha radiation dose of 0.13 rad, what is most nearly the dose equivalent?

(A) 0.13 rem
(B) 0.65 rem
(C) 1.3 rem
(D) 2.6 rem

**58.** An underground piping failure results in the release of 1500 gal of unleaded gasoline into the surrounding soil. A soil vapor extraction system is constructed as the remediation alternative. The chemical of concern has the following characteristics.

| | |
|---|---|
| molecular weight (g/mol) | 106 |
| mole fraction (in gasoline) | 0.018 |
| vapor pressure (atm) | 0.0092 |

The extraction system is designed to provide an air flow rate of 20 ft³/min. The ambient soil temperature is 8°C. If the source is non-diminishing and the vented chemical vapor is emitted directly to the atmosphere without treatment, what is most nearly the mass emission rate of the chemical to the atmosphere?

(A) 0.020 kg/d
(B) 0.25 kg/d
(C) 0.55 kg/d
(D) 0.62 kg/d

**59.** A leaking waste sump at a computer chip fabrication plant has resulted in contaminated groundwater containing the following mix of chemicals.

| mineral spirits | 227 ppb |
| trichloroethene | 289 ppb |
| 1,1,1-trichloroethane | 292 ppb |
| isopropanol | 170 ppb |
| methyl ethyl ketone | 92 ppb |
| isophorone | 86 ppb |
| 1,2-dichloroethene | 183 ppb |
| 1,1-dichloroethane | 46 ppb |

A bench scale isotherm study was conducted using samples of the groundwater. This study provided the following isotherm equation.

$$X/M = 2.837 C_e^{0.431}$$

| $X$ | mass of chemical removed | mg |
| $M$ | mass of GAC capacity consumed | g |
| $C_e$ | chemical concentration at equilibrium | mg/L |

It is expected that the extraction well system will be able to produce about 92 gal/min of continuous flow. Two adsorber vessels, each with 20,000 lbm granular activated carbon (GAC) capacity, are used in a lead-follow configuration. Assume vessels operate at the same pH and temperature for which the isotherm was developed. The required effluent concentration is 0.1 mg/L. What is most nearly the GAC adsorption capacity for the chemical mixture?

(A) 0.59 mg/g
(B) 0.84 mg/g
(C) 1.1 mg/g
(D) 3.3 mg/g

**60.** Arsenic at 270 μg/L has been discovered in a spring serving a rural community with a population of 5000. The spring feeds a small creek that provides the community with a trout fishery. The community has used the spring as a drinking water source for 12 yr, but it is unknown how long the contamination has existed. The toxicology of arsenic is characterized by the following parameters.

| oral route cancer slope factor (CSF) | 1.75 (mg/kg·d)$^{-1}$ |
| bioconcentration factor (BCF) | 44 L/kg |

The exposure factors representing the community are

| ingestion (drinking water) | 2 L/d (adult) |
| ingestion (fish) | 54 g/d |
| lifespan | 70 yr |
| body mass | 70 kg (adult) |

What is most nearly the incremental lifetime cancer risk for an adult who eats fish caught in the creek?

(A) $9.1 \times 10^{-4}$
(B) $2.8 \times 10^{-3}$
(C) $5.3 \times 10^{-3}$
(D) $1.6 \times 10^{-2}$

**61.** A gasoline spill covers an area of 1160 ft² to an average depth of 10.3 ft. The soil in the area consists mostly of medium-to-fine sand. Most nearly, what volume of gasoline was spilled?

(A) 560 gal
(B) 690 gal
(C) 1100 gal
(D) 1800 gal

**62.** Bench tests to evaluate the mass transfer coefficient using an appropriate packing material produced the following results.

| time (s) | concentration (μg/L) |
|---|---|
| 0 | 2000 |
| 20 | 1620 |
| 60 | 1023 |
| 120 | 545 |
| 240 | 134 |
| 360 | 45 |
| 420 | 18 |

Assume a first order reaction stripping factor of 3.0, an air temperature of 25°C, and an atmospheric pressure of 1.0 atm. Assuming a first-order reaction, what is most nearly the value of the mass transfer coefficient?

(A) $0.011 \text{ s}^{-1}$

(B) $0.21 \text{ s·L}/\mu\text{g}$

(C) $4.7 \ \mu\text{g/L·s}$

(D) 18 (unitless)

**63.** The precipitation reaction for lead ($Pb^{+2}$) using sodium hydroxide (NaOH) is

$$Pb^{+2} + 2OH^- \rightarrow Pb(OH)_2$$

The standard free energy values for the reaction are

$$\Delta G \ Pb^{+2} = -5.83 \text{ kcal/mol}$$
$$\Delta G \ OH^- = -36.7 \text{ kcal/mol}$$
$$\Delta G \ Pb(OH)_2 = -108.1 \text{ kcal/mol}$$

Will sodium hydroxide precipitate lead when the lead is present at a concentration of 72 mg/L?

(A) Yes, the standard free energy change, $\Delta G°$, for the reaction with Pb(II) as a reactant is positive.

(B) Yes, the standard free energy change, $\Delta G°$, for the reaction with Pb(II) as a reactant is negative.

(C) No, the standard free energy change, $\Delta G°$, for the reaction with Pb(II) as a reactant is positive.

(D) No, the standard free energy change, $\Delta G°$, for the reaction with Pb(II) as a reactant is negative.

**64.** What is most nearly the intake of a child from age 6 yr to 12 yr through drinking water containing trihalomethanes at a maximum contaminant level (MCL) of 0.080 mg/L?

(A) $1.8 \times 10^{-4}$ mg/kg·d

(B) $3.6 \times 10^{-4}$ mg/kg·d

(C) $1.5 \times 10^{-3}$ mg/kg·d

(D) $3.1 \times 10^{-3}$ mg/kg·d

**65.** An underground storage tank has leaked a light nonaqueous phase liquid (LNAPL) into an unconfined aquifer. The soil-groundwater system has the following characteristics.

| | |
|---|---|
| ambient temperature | 8°C |
| hydraulic conductivity | 19 m/d |
| hydraulic gradient | 0.0017 m/m |
| effective porosity | 0.24 |

What is most nearly the average velocity of the LNAPL if the hydraulic conductivity with respect to the LNAPL is 10 m/d?

(A) 0.0054 m/d

(B) 0.023 m/d

(C) 0.071 m/d

(D) 13 m/d

**66.** Tetrachloroethene (PERC) was found in the soil and groundwater at an abandoned industrial site. The organic carbon content of the soil in the area of contamination is 17%. The organic carbon partition coefficient for PERC is 364 mL/g. What is most nearly the soil-water partition coefficient for PERC in the contaminated soil?

(A) 21 mL/g

(B) 62 mL/g

(C) 340 mL/g

(D) 6200 mL/g

**67.** A workplace exposure survey has produced the following results for benzene.

| | |
|---|---|
| 3 h exposure at concentration (ppm) | 12 |
| 5 h exposure at concentration (ppm) | 4 |
| acceptable ceiling (ppm) | 25 |
| acceptable maximum peak above ceiling concentration (ppm) | 50 |
| duration (min) | 10 |

What is most nearly the time-weighted average exposure to benzene?

(A) 2 ppm

(B) 3 ppm

(C) 7 ppm

(D) 8 ppm

**68.** What is the primary exposure route for radon in drinking water?

(A) Radon gas is emitted from untreated groundwater when used in showers, bathing, and other in-home activities.

(B) Radon gas is emitted from surface water when used in showers, bathing, and other in-home activities.

(C) Particulate radon is ingested by drinking surface water.

(D) Radon is ingested from eating foods grown in radon-containing soil.

**69.** When does the material name on the safety data sheet (SDS) have to exactly match the name printed on the material container?

(A) only when the material will be used in a hospital or other health care facility

(B) only when the material is shipped in containers with capacity greater than 30 L or 25 kg

(C) only when the container is labeled with a NFPA hazard diamond

(D) always

**70.** What physical data is typically NOT included on a safety data sheet (SDS)?

(A) the material appearance and odor

(B) the maximum contaminant level (MCL), recommended MCL (RMCL), and MCL goal (MCLG)

(C) the molecular weight, boiling point, melting point, viscosity, and water solubility for the material

(D) the chemical formula for the material

**71.** The contents of a gas mixture are given.

| gas | gas volume in air (%) |
|---|---|
| ethyl ether | 5.4 |
| ethylene | 7.1 |
| methane | 2.8 |

Most nearly, what is the lower flammability limit (LFL) for the mixture?

(A) 2.3%

(B) 17%

(C) 40%

(D) 63%

**72.** Some substances, when combined, produce different toxicity effects than the same substances produce when alone. How would the toxicity effects from combined exposure to cigarette smoke and asbestos fibers be characterized?

(A) additive

(B) antagonistic

(C) neutral

(D) synergistic

**73.** To reduce noise pollution in the surrounding neighborhood, a manufacturer moves a noisy process to the interior of its site. The distance from the old location to the new location is 661 ft, increasing the distance from the process to the nearest neighbor from 824 ft to 1485 ft. Most nearly, what is the noise reduction experienced by the nearest neighbor? (Assume a point noise source.)

(A) 1.0 dB

(B) 1.9 dB

(C) 2.6 dB

(D) 5.1 dB

**74.** Which of the following performance standards is NOT included among the risk-based performance standards for chemical facilities as established by the U.S. Department of Homeland Security?

(A) deterrence, detection, and delay

(B) resistance and defense

(C) restriction of area perimeter

(D) shipping, receipt, and storage

**75.** Suppressed free discharge flow occurs over a 1.6 m rectangular weir. The height of the water over the weir is measured at 0.23 m. Most nearly, what is the water flow rate over the weir?

(A) 0.25 m³/s

(B) 0.32 m³/s

(C) 0.59 m³/s

(D) 1.1 m³/s

**76.** A municipality operates an anaerobic digester. To increase the value of the digester gases and reduce environmental impacts, the city is considering separating carbon dioxide for recovery from other digester gases and then selling it. The Henry's constant for carbon dioxide is 1040 atm/mol fraction at 10°C. Most nearly how many 1 L bottles of beverage per 100 kg of recovered carbon dioxide can be carbonated at 10°C and 1.2 atm?

(A) 3000 L/100 kg

(B) 6100 L/100 kg

(C) 25 000 L/100 kg

(D) 35 000 L/100 kg

**77.** A hydrologic basin is characterized by a historical rainfall record that is normally distributed with a mean of 48 in, a standard deviation of 11.2 in, and a skew coefficient of 0.09. Most nearly, what is the expected rainfall depth for a 15-year return period for the basin?

(A) 50 in

(B) 59 in

(C) 65 in

(D) 69 in

**78.** How is float determined when applying the critical path method for project scheduling?

(A) early finish–early start (EF–ES)

(B) early finish–late start (EF–LS)

(C) early start–late finish (ES–LF)

(D) late start–early start (LS–ES)

**79.** A municipal project has an initial cost of $2.3M and a 10-year term, financed at an annual interest rate of 1.5%. The project assets will have $143,000 of salvage value at the end of the project term. The annual maintenance costs are shown.

| year | 1 | 2 | 3 | 4 | 5 | 6 | 7 | 8 | 9 | 10 |
|---|---|---|---|---|---|---|---|---|---|---|
| cost ($10³) | 18 | 13 | 14 | 15 | 16 | 17 | 18 | 19 | 20 | 21 |

Most nearly, how much money should be borrowed to finance the project over its full term?

(A) $2,600,000

(B) $2,630,000

(C) $2,670,000

(D) $2,700,000

**80.** The Global Positioning System (GPS) is a satellite constellation that makes it possible to receive coordinates of a location by triangulating from multiple satellites. Which datum is used by the GPS constellation to provide $x$, $y$, and $z$ coordinates?

(A) WGS84 UTM

(B) WGS84 Geographic

(C) NAD27 UTM

(D) NAD83 UTM

# STOP!

### DO NOT CONTINUE!

This concludes the Afternoon Session of the examination. If you finish early, check your work and make sure that you have followed all instructions. After checking your answers, submit your solutions and leave the examination room. Once your answers are submitted you will not be able to access them again.

# Morning Session Answer Sheet 2

| | | | |
|---|---|---|---|
| 81. Ⓐ Ⓑ Ⓒ Ⓓ | 91. Ⓐ Ⓑ Ⓒ Ⓓ | 101. Ⓐ Ⓑ Ⓒ Ⓓ | 111. Ⓐ Ⓑ Ⓒ Ⓓ |
| 82. Ⓐ Ⓑ Ⓒ Ⓓ | 92. Ⓐ Ⓑ Ⓒ Ⓓ | 102. Ⓐ Ⓑ Ⓒ Ⓓ | 112. Ⓐ Ⓑ Ⓒ Ⓓ |
| 83. Ⓐ Ⓑ Ⓒ Ⓓ | 93. Ⓐ Ⓑ Ⓒ Ⓓ | 103. Ⓐ Ⓑ Ⓒ Ⓓ | 113. Ⓐ Ⓑ Ⓒ Ⓓ |
| 84. Ⓐ Ⓑ Ⓒ Ⓓ | 94. Ⓐ Ⓑ Ⓒ Ⓓ | 104. Ⓐ Ⓑ Ⓒ Ⓓ | 114. Ⓐ Ⓑ Ⓒ Ⓓ |
| 85. Ⓐ Ⓑ Ⓒ Ⓓ | 95. Ⓐ Ⓑ Ⓒ Ⓓ | 105. Ⓐ Ⓑ Ⓒ Ⓓ | 115. Ⓐ Ⓑ Ⓒ Ⓓ |
| 86. Ⓐ Ⓑ Ⓒ Ⓓ | 96. Ⓐ Ⓑ Ⓒ Ⓓ | 106. Ⓐ Ⓑ Ⓒ Ⓓ | 116. Ⓐ Ⓑ Ⓒ Ⓓ |
| 87. Ⓐ Ⓑ Ⓒ Ⓓ | 97. Ⓐ Ⓑ Ⓒ Ⓓ | 107. Ⓐ Ⓑ Ⓒ Ⓓ | 117. Ⓐ Ⓑ Ⓒ Ⓓ |
| 88. Ⓐ Ⓑ Ⓒ Ⓓ | 98. Ⓐ Ⓑ Ⓒ Ⓓ | 108. Ⓐ Ⓑ Ⓒ Ⓓ | 118. Ⓐ Ⓑ Ⓒ Ⓓ |
| 89. Ⓐ Ⓑ Ⓒ Ⓓ | 99. Ⓐ Ⓑ Ⓒ Ⓓ | 109. Ⓐ Ⓑ Ⓒ Ⓓ | 119. Ⓐ Ⓑ Ⓒ Ⓓ |
| 90. Ⓐ Ⓑ Ⓒ Ⓓ | 100. Ⓐ Ⓑ Ⓒ Ⓓ | 110. Ⓐ Ⓑ Ⓒ Ⓓ | 120. Ⓐ Ⓑ Ⓒ Ⓓ |

# Morning Session 2

**81.** A municipal sanitary sewer authority is experiencing infiltration and inflow (I/I) problems. Monitoring at manholes has produced the following I/I data for a section of the sewer system.

| pipe no. | up manhole | down manhole | pipe diameter (mm) | pipe length (m) | average flow (L/d) |
|---|---|---|---|---|---|
| 1 | 3 | 2 | 600 | 85 | 2500 |
| 2 | 2 | 7 | 600 | 46 | 2000 |
| 3 | 4 | 3 | 300 | 40 | 140 |
| 4 | 5 | 4 | 250 | 67 | 500 |
| 5 | 8 | 6 | 200 | 91 | 50 |
| 6 | 7 | 8 | 200 | 131 | 80 |
| 7 | 1 | 6 | 150 | 50 | 110 |

The greatest benefit per cost unit would result from rehabilitating which lines?

(A) 2

(B) 1 and 2

(C) 1, 2, and 4

(D) 1, 2, 4, and 6

**82.** What is most nearly the impact on the incremental annual BOD mass loading rate to a wastewater treatment plant if conventional plumbing fixtures and appliances are replaced with water conservation plumbing fixtures and appliances?

(A) no change because the concentration will increase

(B) increases because the concentration will increase

(C) decreases because the volume will decrease

(D) decreases because the volume and concentration will decrease

**83.** The 25 yr storm produces 2 in of rainfall. The storm distribution is type II, as represented by the following illustration.

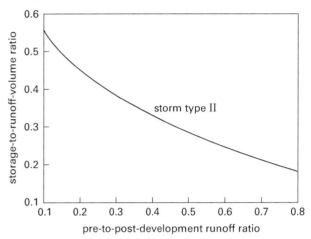

What is most nearly the required ratio of storage to runoff volume when the ratio of pre-to-post-development runoff is 0.42?

(A) 0.29

(B) 0.32

(C) 0.38

(D) 0.48

**84.** Type II settling column tests were performed for a flocculated suspension. The settling column tests provided the following results.

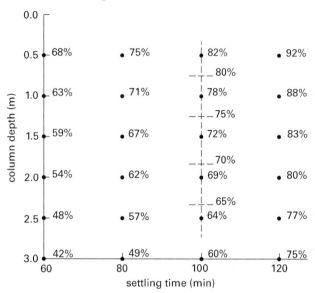

What is the effect on efficiency if the settling time is increased to 120 min?

(A) decreases

(B) increases

(C) remains unchanged if the settling depth remains unchanged

(D) remains unchanged if the overflow rate remains unchanged

**85.** Four parallel sedimentation basins will treat a total flow of 12 000 m³/d containing a total suspended solids (TSS) concentration of 195 mg/L. Assuming 80% efficiency, what is most nearly the daily volume of TSS removed if the solids content is 18%?

(A) 0.50 m³/d

(B) 2.3 m³/d

(C) 10 m³/d

(D) 16 m³/d

**86.** An electroplater produces wastewater at a continuous rate of 1.8 m³/min with $Cr^{+6}$ (measured as $CrO_3$) at a concentration of 534 mg/L. Sodium bisulfite ($NaHSO_3$) and sulfuric acid ($H_2SO_4$) are selected to reduce the chromium from the hexavalent to the trivalent form. The chemical equation for the reduction reaction is

$$4CrO_3 + 6NaHSO_3 + 3H_2SO_4 \rightarrow 3Na_2SO_4 + 2Cr_2(SO_4)_3 + 6H_2O$$

What is most nearly the annual mass of reduced chromium that will require precipitation, assuming the facility operates 24 h/d, 365 d/yr?

(A) $1.3 \times 10^5$ kg/yr

(B) $5.1 \times 10^5$ kg/yr

(C) $9.9 \times 10^5$ kg/yr

(D) $2.0 \times 10^6$ kg/yr

**87.** What is most nearly the daily volume of sludge wasted from a secondary clarifier at 6% solids corresponding to a sludge production rate of 500 kg/d?

(A) 5.4 m³/d

(B) 8.3 m³/d

(C) 19 m³/d

(D) 27 m³/d

**88.** A complete mix activated sludge process has been selected for treatment of a wastewater. Available design information is presented in the following table.

| parameter | value |
|---|---|
| flow rate from primary clarifiers | 5000 m³/d |
| solids recycle flow rate | 180 m³/d |
| mixed liquor volatile suspended solids, MLVSS | 2300 mg/L |
| recycle solids | 10 000 mg/L |
| influent substrate | 192 mg/L $BOD_5$ |
| effluent substrate | 20 mg/L soluble $BOD_5$ |
| yield coefficient | 0.5 at 20°C |
| growth rate coefficient | 0.40/d at 20°C |
| decay coefficient | 0.05/d at 20°C |
| minimum mean cell residence time | 3.0 d |
| activated sludge safety factor | 2.5 |

What is most nearly the hydraulic residence time in the bioreactor?

(A) 4.9 h

(B) 5.5 h

(C) 6.7 h

(D) 7.5 h

**89.** A biological treatment process produces an effluent that requires clarification. The clarifier solids flux and settling velocity are 2.3 kg/m²·h and 1.34 m/h, respectively. What is most nearly the required overflow rate for the clarifier if settling velocity controls design?

(A) 1.3 m³/m²·h

(B) 1.7 m³/m²·h

(C) 3.0 m³/m²·h

(D) 9.4 m³/m²·h

**90.** An aeration blower must provide an air flow of 100 ft³/sec for wastewater treatment. The ambient temperature of the air is 73°F, and the ambient pressure is 1.0 atm. The maximum allowable outlet pressure is 1.3 atm. Most nearly, what is the power required to achieve the necessary air flow at an operating efficiency of 80%?

(A) 50 hp

(B) 110 hp

(C) 140 hp

(D) 260 hp

**91.** Results of dye tracer studies that were conducted to define the flow characteristics of a reaction tank for a water process are tabulated and graphed as shown. The reactor volume was 25,000 gal and the flow rate to the reactor was 500 gal/min.

| time (min) | concentration (μg/L) |
|---|---|
| 20 | 0 |
| 30 | 100 |
| 40 | 390 |
| 50 | 148 |
| 60 | 83 |
| 70 | 47 |
| 80 | 25 |
| 90 | 12 |
| 100 | 7.5 |
| 110 | 2.5 |

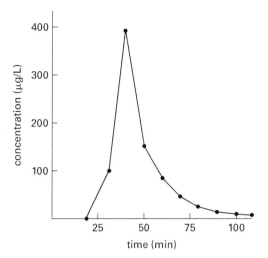

What is most nearly the reactor mean hydraulic residence time?

(A) 20 min

(B) 47 min

(C) 65 min

(D) 110 min

**92.** A tank-impeller-type flash mixer is needed to treat a design flow of 5000 m³/d. The velocity gradient is 850 s⁻¹ with a hydraulic residence time of 120 s. What is most nearly the total power required at the flash mixer, assuming a single mixing tank is used?

(A) 0.35 kW

(B) 5.0 kW

(C) 17 kW

(D) 30 kW

**93.** A hydrologic basin is characterized by fair condition grassland with a curve number (CN) of 49. For a storm event producing 3.6 in of precipitation, most nearly what runoff depth results?

(A) 0.13 in

(B) 0.19 in

(C) 0.95 in

(D) 2.7 in

**94.** A 100 ha drainage basin is characterized by a 2 yr rainfall intensity of 5.3 cm/h. What is most nearly the peak discharge at the 2 yr storm? Use the rational equation method.

(A) 0.5 m³/s
(B) 2.2 m³/s
(C) 3.7 m³/s
(D) 6.3 m³/s

**95.** A single construction Class 3 structure has an effective area of 9200 ft², an occupancy coefficient of 0.8, and a combined exposure and communication factor of 0.23. Most nearly, what is the needed fire flow for the structure?

(A) 570 gpm
(B) 1400 gpm
(C) 1500 gpm
(D) 1700 gpm

**96.** An industrial discharger is unable to satisfy the permit pre-treatment requirements for discharge to the local municipal wastewater treatment plant. The current wastewater characteristics generated by the discharger are 2000 mg/L $BOD_5$ at 400 m³/d. The pre-treatment system consists of an anaerobic pond based on a volumetric mass loading of 0.35 kg BOD/m³·d. Most nearly, what minimum liquid depth should be maintained in the pond to satisfy volumetric mass loading criteria?

(A) 1.5 m
(B) 2.3 m
(C) 3.0 m
(D) 7.6 m

**97.** A wastewater is treated using a complete mix activated sludge process followed by clarification. National Pollutant Discharge Elimination System (NPDES) permit requirements limit discharges of volatile suspended solids (VSS) to 30 mg/L and total phosphorous to 1.0 mg/L. The plant routinely violates its discharge limits with VSS averaging 60 mg/L and total phosphorous averaging 1.3 mg/L. The source of the violations is believed to be associated with the operation of the two secondary clarifiers. Which of the following actions would NOT result in improved performance of the secondary clarifiers?

(A) constructing a third secondary clarifier
(B) feeding chemicals to improve settling of the biofloc
(C) decreasing the depth of the inlet baffles
(D) increasing the length of the effluent weirs

**98.** Water samples characterized by the following analysis were collected from a well intended to provide 500 gal/min flow to a manufacturing facility.

| ion | concentration (mg/L) |
|---|---|
| $SO_4^{-2}$ | 28 |
| $Ca^{+2}$ | 67 |
| $HCO_3^-$ | 353 |

Neglecting excess alkalinity, what is most nearly the daily sludge volume at 35% solids generated from the lime-soda ash softening the water?

(A) 0.32 m³/d
(B) 1.3 m³/d
(C) 2.6 m³/d
(D) 4.5 m³/d

**99.** A sedimentation basin is required to treat 10 000 m³/d of flow containing 234 mg/L total suspended solids (TSS). Assuming 80% efficiency and 30% solids, what is most nearly the daily volume of sludge requiring disposal?

(A) 0.56 m³/d
(B) 1.9 m³/d
(C) 6.2 m³/d
(D) 7.8 m³/d

**100.** The average daily water use characteristics of a community are presented in the following illustration.

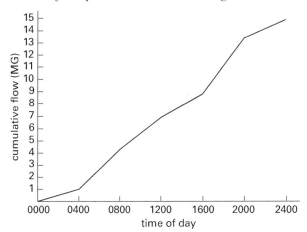

What is most nearly the daily demand?

(A) 0.63 MGD

(B) 2.5 MGD

(C) 11 MGD

(D) 15 MGD

**101.** A 200 mL water sample was titrated with 0.03N sulfuric acid ($H_2SO_4$). The initial pH was 7.8 and 28 mL of acid was added to reach pH 4.5. What is most nearly the total alkalinity of the sample expressed in mg/L as $CaCO_3$?

(A) 110 mg/L as $CaCO_3$

(B) 140 mg/L as $CaCO_3$

(C) 210 mg/L as $CaCO_3$

(D) 420 mg/L as $CaCO_3$

**102.** Water samples were collected and submitted to a commercial analytical laboratory for analysis. The laboratory provided the following results.

| ion | concentration (mg/L) |
|---|---|
| $SO_4^{-2}$ | 17 |
| $Na^+$ | 48 |
| $K^+$ | 14 |
| $Cl^-$ | 39 |
| $Ca^{+2}$ | 56 |
| $Mg^{+2}$ | 12 |
| $HCO_3^-$ | 237 |

What is most nearly the noncarbonate hardness concentration when both the total hardness and the carbonate hardness are each 100 mg/L as $CaCO_3$?

(A) 0 mg/L as $CaCO_3$

(B) 50 mg/L as $CaCO_3$

(C) 100 mg/L as $CaCO_3$

(D) 200 mg/L as $CaCO_3$

**103.** The illustration shown presents a breakpoint chlorination curve for a water sample.

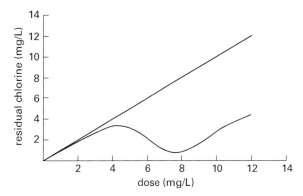

What is most nearly the minimum chlorine dose required to obtain a free chlorine residual?

(A) 1.5 mg/L

(B) 4.3 mg/L

(C) 7.7 mg/L

(D) 9.7 mg/L

**104.** A pond occupies an area of 24 ac to an average depth of 8 ft. The creek feeding the pond has an average flow of 2.23 $ft^3$/min with a total Kjeldahl nitrogen (TKN) concentration of 1.43 mg/L as N. What is most nearly the annual TKN loading to the pond?

(A) 0.052 lbm/ac-ft-yr

(B) 0.54 lbm/ac-ft-yr

(C) 3.6 lbm/ac-ft-yr

(D) 19 lbm/ac-ft-yr

**105.** Which of the following are important factors for determining incinerator efficiency?

(A) burn temperature, oxygen level, turbulence, residence time

(B) burn temperature, viscosity, turbulence, residence time

(C) burn temperature, oxygen level, corrosivity, residence time

(D) burn temperature, oxygen level, flame velocity, turbulence

**106.** The ambient water vapor pressure is 0.089 psi at a temperature of 36°F. Most nearly, what is the relative humidity?

(A) 10%
(B) 15%
(C) 85%
(D) 90%

**107.** An electrostatic precipitator (ESP) is being considered for use as an air pollution control process. The air flow rate requiring treatment is 14 m³/s. For a drift velocity of 1.0 m/s, what is most nearly the plate area required for the ESP if an 80% efficiency is desired?

(A) 2 m²
(B) 10 m²
(C) 23 m²
(D) 70 m²

**108.** Particulate matter with a mean diameter of 2.5 μm and a density of 0.58 g/cm³ is present in air at 1.3 atm and 30°C. The air density at 1.3 atm is 0.00152 g/cm³ and the air absolute viscosity is $1.9 \times 10^5$ kg/m·s. What is most nearly the particle-settling velocity?

(A) 0.0011 cm/s
(B) 0.010 cm/s
(C) 0.10 cm/s
(D) 0.42 cm/s

**109.** Incineration of an organic hazardous contaminant requires a destruction and removal efficiency of 99.99%. The unit mass feed rate is 1000 kg/h. Most nearly, what is the maximum allowable mass emission rate?

(A) 0.010 kg/h
(B) 0.10 kg/h
(C) 0.90 kg/h
(D) 1.0 kg/h

**110.** In the geographic location covered by the illustration, approximately what percent of the time does the wind speed blowing from W-SW exceed 10 mph?

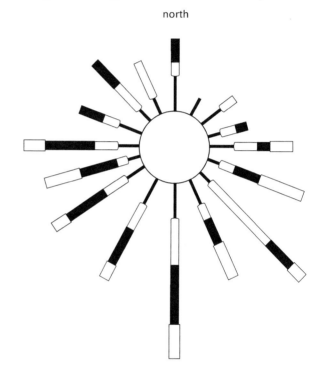

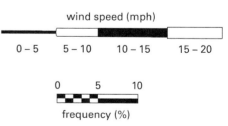

(A) 2%
(B) 5%
(C) 9%
(D) 12%

**111.** What is the traditional approach employed by the Environmental Protection Agency (EPA) for air pollution control?

(A) command-and-control based on uniform emission standards

(B) geographic-specific based on industrial category

(C) variable emission standards based on climatic conditions and topography

(D) site-by-site based on economic and demographic factors

**112.** What are the primary nuisance components of photochemical smog?

(A) particulates and sulfur oxides

(B) carbon dioxide and CFCs

(C) ozone and aldehydes

(D) nitrogen oxides and carbon monoxide

**113.** Octane ($C_8H_{18}$) has a specific gravity of 0.701. Most nearly, what mass of carbon dioxide is emitted to the atmosphere from the combustion of 100 gal of octane?

(A) 33 kg

(B) 260 kg

(C) 820 kg

(D) 2100 kg

**114.** How will increasing the cross-sectional area influence the efficiency of a cross current-flow scrubber for air pollution control?

(A) The efficiency will increase because the scrubber volume will increase.

(B) The efficiency will increase because the gas velocity, $v_g$, will increase.

(C) The efficiency will decrease because the gas velocity, $v_g$, will decrease.

(D) The efficiency will decrease because the liquid-gas flow ratio will decrease.

**115.** A stack with an effective height of 120 m emits pollutants at 5.4 kg/s. Wind speed measured 10 m above the ground surface is 5.3 m/s on a cloudy day with slight incoming solar radiation. Most nearly, how far downwind is the maximum ground-level concentration of pollutants emitted from the stack observed?

(A) 210 m

(B) 400 m

(C) 800 m

(D) 2500 m

**116.** What condition defines a non-attainment area?

(A) inability of a region to meet the National Emission Standards for Hazardous Air Pollutants (NESHAP)

(B) inability of a region to meet the National Ambient Air Quality Standards (NAAQS)

(C) inability of a region to meet New Source Performance Standards (NSPS)

(D) none of the above

**117.** A baghouse is used to remove particulate from a waste air stream. The baghouse is designed for a bag diameter of 15 cm and bag height of 2.5 m. Most nearly, how many bags are required per 100 m² of total fabric area?

(A) 85

(B) 170

(C) 530

(D) 850

**118.** A cyclone separator is used to remove particulate from a contaminated air stream. The cyclone separator characteristics are

| | |
|---|---|
| cyclone body diameter | 4.5 ft |
| cyclone inlet height | $0.15D$ |
| cyclone body length | $1.5D$ |
| cyclone cone length | $2.5D$ |

What is most nearly the number of effective turns the air makes in the cyclone?

(A) 11

(B) 18

(C) 27

(D) 83

**119.** The following illustrations present atmospheric stability conditions. For illustration I, the lapse rate is $-0.0159°C/m$.

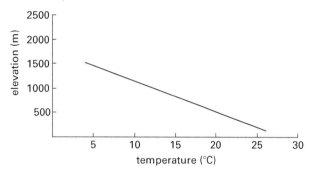

For illustration II, the lapse rate is $-0.0064°C/m$.

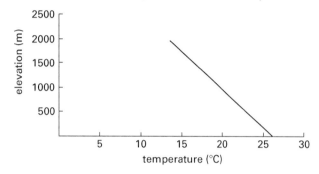

For illustration III, the lapse rate is $-0.0098°C/m$.

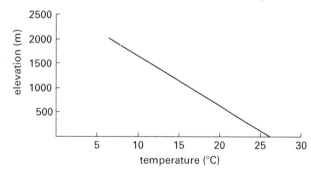

For illustration IV, the lapse rate is $0.0043°C/m$.

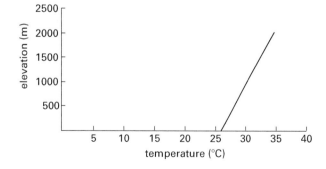

Which illustration depicts neutral conditions?

(A) illustration I

(B) illustration II

(C) illustration III

(D) illustration IV

**120.** A pollutant is emitted from a source at 26 kg/s on a day when the wind speed is 3.2 m/s and the atmospheric stability is category C. What is most nearly the ground-level concentration at 3800 m along the plume centerline for a release from a 110 m stack?

(A) $22 \text{ mg/m}^3$

(B) $29 \text{ mg/m}^3$

(C) $33 \text{ mg/m}^3$

(D) $38 \text{ mg/m}^3$

# STOP!

### DO NOT CONTINUE!

This concludes the Morning Session of the examination. If you finish early, check your work and make sure that you have followed all instructions. After checking your answers, submit your solutions and leave the examination room. Once your answers are submitted you will not be able to access them again.

# Afternoon Session Answer Sheet 2

| | | | |
|---|---|---|---|
| 121. (A) (B) (C) (D) | 131. (A) (B) (C) (D) | 141. (A) (B) (C) (D) | 151. (A) (B) (C) (D) |
| 122. (A) (B) (C) (D) | 132. (A) (B) (C) (D) | 142. (A) (B) (C) (D) | 152. (A) (B) (C) (D) |
| 123. (A) (B) (C) (D) | 133. (A) (B) (C) (D) | 143. (A) (B) (C) (D) | 153. (A) (B) (C) (D) |
| 124. (A) (B) (C) (D) | 134. (A) (B) (C) (D) | 144. (A) (B) (C) (D) | 154. (A) (B) (C) (D) |
| 125. (A) (B) (C) (D) | 135. (A) (B) (C) (D) | 145. (A) (B) (C) (D) | 155. (A) (B) (C) (D) |
| 126. (A) (B) (C) (D) | 136. (A) (B) (C) (D) | 146. (A) (B) (C) (D) | 156. (A) (B) (C) (D) |
| 127. (A) (B) (C) (D) | 137. (A) (B) (C) (D) | 147. (A) (B) (C) (D) | 157. (A) (B) (C) (D) |
| 128. (A) (B) (C) (D) | 138. (A) (B) (C) (D) | 148. (A) (B) (C) (D) | 158. (A) (B) (C) (D) |
| 129. (A) (B) (C) (D) | 139. (A) (B) (C) (D) | 149. (A) (B) (C) (D) | 159. (A) (B) (C) (D) |
| 130. (A) (B) (C) (D) | 140. (A) (B) (C) (D) | 150. (A) (B) (C) (D) | 160. (A) (B) (C) (D) |

# Afternoon Session 2

**121.** An industry discharges 300,000 L/d of a wastewater containing 2.1 mol hydrochloric acid. Assuming a 1-to-1 molar ratio between the acid and the base, what is most nearly the daily mass of sodium hydroxide required to neutralize the acid?

(A) 630 kg/d
(B) 11 000 kg/d
(C) 14 000 kg/d
(D) 25 000 kg/d

**122.** Solid waste generated by a municipality is characterized as follows.

| waste component | mass (%) | moisture (%) |
|---|---|---|
| food | 13 | 70 |
| glass | 6 | 2 |
| plastic | 4 | 2 |
| paper | 37 | 6 |
| cardboard | 10 | 5 |
| textiles | 1 | 10 |
| ferrous metal | 8 | 3 |
| nonferrous metal | 2 | 2 |
| wood | 4 | 20 |
| yard clippings | 15 | 60 |

What is most nearly the discarded moisture content of the bulk waste?

(A) 18%
(B) 22%
(C) 36%
(D) 44%

**123.** What are the primary constituents of the landfill gas?

(A) $CH_4$
(B) $CH_4$ and $CO_2$
(C) $CH_4$, $CO_2$, and $NH_3$
(D) $CH_4$, $CO_2$, $NH_3$, and $H_2S$

**124.** An onshore facility uses an above-ground tank farm for bulk storage of fuel and lubricating oils. The tank farm, covered by a current spill prevention, control, and countermeasure plan, includes the following tanks.

| tank no. | product stored | tank volume (gal) |
|---|---|---|
| 1 | no. 2 fuel oil | 1,000,000 |
| 2 | no. 2 fuel oil | 1,000,000 |
| 3 | no. 2 fuel oil | 410,000 |
| 4 | no. 2 fuel oil | 650,000 |
| 5 | SAE 60W lubricating oil | 55,000 |
| 6 | SAE 75W lubricating oil | 18,000 |
| 7 | SAE 90W lubricating oil | 27,000 |

The bermed containment area surrounding the tanks must accommodate 110% of the volume of the largest tank plus the volume of rainfall produced by the 25 yr return, 24 hr rainfall depth of 14 in. The tank farm occupies 1.85 ac. If a berm were used for containment, what would most nearly be the required berm height assuming 1.5 ft freeboard?

(A) 2.9 ft
(B) 3.3 ft
(C) 4.5 ft
(D) 5.2 ft

**125.** A manufacturer operates an activated sludge plant for production wastewaters. The activated sludge plant produces 4 m³/d of wasted sludge dewatered to 76% moisture that requires disposal. The heating value of the wasted sludge is 30 000 kJ/kg with a 5% ash content. What is most nearly the hourly heat value available from the dry sludge solids?

(A) $3.0 \times 10^4$ kJ/h
(B) $1.2 \times 10^6$ kJ/h
(C) $3.9 \times 10^6$ kJ/h
(D) $9.0 \times 10^6$ kJ/h

**126.** A 500 mm thick landfill has a hydraulic conductivity of $10^{-9}$ m/s and a porosity of 0.2. Leachate is expected to accumulate to a depth of 140 mm above the liner. Most nearly, what is the breakthrough time for the leachate to penetrate the liner?

(A) 1.6 yr
(B) 2.6 yr
(C) 3.2 yr
(D) 4.9 yr

**127.** Hydrogen cyanide (HCN) gas is recovered from an air stream using a countercurrent scrubber operated under vacuum. The gas flow rate through the scrubber is 1.9 m³/s with a hydrogen cyanide concentration of 1.94 g/m³. The scrubbing solution is 0.1 N sodium hydroxide (NaOH). What is most nearly the required flow rate for the sodium hydroxide solution?

(A) 22 L/min
(B) 41 L/min
(C) 55 L/min
(D) 82 L/min

**128.** A community generates solid waste with the following characteristics.

| component | mass (%) | moisture (%) | discarded density (kg/m³) | discarded energy (kJ/kg) | ash (%) |
|---|---|---|---|---|---|
| food | 16 | 70 | 290 | 4650 | 5 |
| glass | 6 | 2 | 195 | 150 | 98 |
| plastic | 5 | 2 | 65 | 32 600 | 10 |
| paper | 44 | 6 | 85 | 16 750 | 6 |
| ferrous metal | 5 | 3 | 320 | 840 | 98 |
| nonferrous metal | 5 | 2 | 160 | 700 | 96 |
| yard clippings | 19 | 60 | 105 | 6500 | 4.5 |

What is most nearly the energy content of the discarded bulk waste?

(A) 1700 kJ/kg
(B) 9300 kJ/kg
(C) 11 000 kJ/kg
(D) 13 000 kJ/kg

**129.** Stepwise formation constants and associated chemical equations for lead hydroxide are

$$Pb^{+2} + OH^- \rightarrow PbOH^+$$

$$k_1 = \frac{[PbOH]^+}{[Pb^{+2}][OH^-]} = 10^{7.82}$$

$$PbOH^+ + OH^- = Pb(OH)_2$$

$$k_2 = \frac{[Pb(OH)_2]}{[PbOH^+][OH^-]} = 10^{3.03}$$

$$Pb(OH)_2 + OH^- \rightarrow Pb(OH)_3$$

$$k_3 = \frac{[Pb(OH)_3^-]}{[Pb(OH)_2][OH^-]} = 10^{3.73}$$

If present as a concentration of 72 mg/L, what is the dominant Pb(II) species at pH 7.6?

(A) $Pb^{+2}$
(B) $PbOH^+$
(C) $Pb(OH)_2$
(D) $Pb(OH)_3^-$

**130.** The results of a time study and route analysis for curbside residential waste collection are as follows.

| | |
|---|---|
| population | 1400 |
| solid waste generation rate | 3.6 lbm/day·person |
| number of residences | 388 |
| average driving time between residences | 18 sec |
| average pick-up/load time at each residence | 32 sec |
| travel time from truck yard to route start | 28 min |
| average travel time between route and landfill | 47 min |
| time to unload at landfill | 15 min |
| travel time from landfill to truck yard | 34 min |
| truck compacted waste capacity | 10 yd³ |
| truck compaction ratio | 3.1 |
| typical waste as-discarded density | 240 lbm/yd³ |

Most nearly, how many days are required for one crew to collect all the waste generated by the community in one week?

(A) 1 day
(B) 2 days
(C) 3 days
(D) 4 days

**131.** A municipal solid waste collected from a population of 117,000 people includes 24% recycled materials that are sorted from the waste at a processing/transfer station. The population generates waste at about 3.9 lbm/day, and the waste that is not recycled is landfilled. The landfilled waste is compacted to 1100 lbm/yd$^3$ in a landfill with a total design capacity of 3,170,000 yd$^3$. What is most nearly the maximum soil-cover-to-waste ratio permissible if the desired landfill life is 25 yr?

(A) 1:1.1
(B) 1:2.5
(C) 1:5.3
(D) 1:9.8

**132.** A wastewater sludge is dewatered to 34% solids with a specific gravity of 1.16. For each 1 m$^3$ of wet sludge volume, the mass of dry solids is most nearly

(A) 120 kg
(B) 350 kg
(C) 400 kg
(D) 750 kg

**133.** What parameter is used to evaluate the bioconcentration of a toxic contaminant in fish tissue?

(A) chronic daily intake (CDI)
(B) reference dose (RfD)
(C) bioconcentration factor (BCF)
(D) acceptable daily intake (ADI)

**134.** A workplace exposure survey has produced the following results.

| | acetone | sec-butanol |
|---|---|---|
| 1 h exposure at concentration (ppm) | 1080 | 180 |
| 2 h exposure at concentration (ppm) | 630 | 85 |
| 5 h exposure at concentration (ppm) | 470 | 15 |
| 8 h TWA PEL* (ppm) | 1000 | 150 |

*TWA PEL is the time-weighted average peak exposure limit.

What is most nearly the cumulative time-weighted average exposure to the acetone and sec-butanol mixture?

(A) 0.56
(B) 0.94
(C) 1.8
(D) 640

**135.** Efficiencies for components of a ventilation system are shown.

| component | efficiency |
|---|---|
| fan | 0.93 |
| transmission | 0.86 |
| motor | 0.89 |
| controller | 0.95 |

Most nearly, what is the overall efficiency of the system?

(A) 0.68
(B) 0.86
(C) 0.91
(D) 0.95

**136.** An instrument used to measure gamma radiation detects 0.016 R/min at a distance from the source of 2 m. Most nearly, what should the instrument measure at 10 m from the source if the inverse-square law is used?

(A) 0.010 mR/min
(B) 0.64 mR/min
(C) 3.2 mR/min
(D) 7.2 mR/min

**137.** An underground storage tank piping failure has resulted in the release of unleaded gasoline into the surrounding soil. Site conditions will not allow excavation of the tank or excavation of the contaminated soil. A soil vapor extraction system has been constructed as the remediation alternative. The gasoline contains the following chemicals of concern.

| | chemical | | | |
|---|---|---|---|---|
| | A | B | C | D |
| weight fraction in gasoline (g/g) | 0.018 | 0.082 | 0.021 | 0.029 |
| molecular weight (g/mol) | 78 | 92 | 106 | 106 |
| mole fraction | 0.020 | 0.091 | 0.018 | 0.016 |
| vapor pressure (atm) | 0.10 | 0.029 | 0.0092 | 0.0076 |
| Henry's constant (atm·m$^3$/mol) | 0.0055 | 0.0064 | 0.0084 | 0.0069 |
| MCL ($\mu$g/L) | 5 | 1000 | 700 | 10 000 |

Which of the four chemicals of concern will likely control design?

(A) chemical A
(B) chemical B
(C) chemical C
(D) chemical D

**138.** Bulk quantities of trichloroethane and sodium hydroxide are stored in the same space. If the two chemicals are allowed to mix, what would be the consequences?

(A) can mix without consequences
(B) heat and flammable gas generation
(C) heat and toxic gas generation with fire
(D) may be hazardous, but unknown

**139.** A waste has the following characteristics.

| characteristic | waste |
|---|---|
| ignitability (flash point, °C) | 92 |
| corrosivity (pH) | 12.7 |
| reactivity (reactive, yes/no) | yes |
| toxicity (contaminant, concentration, $\mu$g/L) | — |

If simple neutralization was applied as treatment for the waste, would it still meet criteria for classification as hazardous waste?

(A) Yes, because it would still be hazardous based on ignitibility.
(B) Yes, because it would still be hazardous based on toxicity.
(C) No, because it was not hazardous based on corrosivity before neutralization.
(D) No, because it was hazardous based on corrosivity only.

**140.** Radiation intensity measured 1.8 ft from a radiation source is 162,000 mR/hr. Most nearly, what distance from the source is required to experience an intensity of 100 mR/hr?

(A) 32 ft
(B) 45 ft
(C) 54 ft
(D) 72 ft

**141.** A pumping test was performed for a well screened in a 8.2 ft thick confined aquifer. At steady state, the drawdown at 7.3 ft from the well casing was 5.7 ft and at 14.8 ft from the well casing was 9.1 ft at a pumping rate of 41 gpm. Most nearly, what is the hydraulic conductivity of the aquifer?

(A) 9.6 ft/day
(B) 29 ft/day
(C) 32 ft/day
(D) 39 ft/day

**142.** Ethylbenzene is present in an aquifer at 219 ppb. The aquifer soil is characterized by an organic carbon fraction of 0.32. Most nearly, what is the soil-water partition coefficient for ethylbenzene in the aquifer?

(A) 27 mL/g
(B) 350 mL/g
(C) 1100 mL/g
(D) 3400 mL/g

**143.** A chemical has the characteristics shown.

| | |
|---|---|
| molecular weight | 120 g/mol |
| partial pressure | 0.26 atm at 25°C |
| solubility in water | 9300 mg/L at 25°C |

What is most nearly the value of the unitless Henry's constant?

(A) 0.014

(B) 0.14

(C) 0.19

(D) 0.34

**144.** Granular activated carbon (GAC) was selected to remove an organic solvent from groundwater. An extraction well system will be able to produce about 500 m³/d of continuous flow and will result in a GAC use rate of 280 kg/d. Two adsorber vessels of 10 000 kg GAC capacity each are used in a lead-follow configuration. The vessel bed volume is 20 m³. What is most nearly the adsorption vessel empty bed contact time (EBCT)?

(A) 29 min

(B) 58 min

(C) 87 min

(D) 120 min

**145.** An underground storage tank has leaked a light nonaqueous phase liquid (LNAPL) into an unconfined aquifer. The LNAPL has a density of 0.811 g/cm³ and kinematic viscosity of 8.32 mm²/s. If the intrinsic permeability of the soil is $1.0 \times 10^{-10}$ m², what is most nearly the hydraulic conductivity with respect to the fuel?

(A) 0.020 m/d

(B) 0.27 m/d

(C) 3.2 m/d

(D) 10 m/d

**146.** The following chemical reaction shows the oxidation of one compound to form another compound.

$$CH_3CH_2CHOHCH_2CH_3 + \tfrac{1}{2}O_2 \rightarrow H_2O + CH_3CH_2COCH_2CH_3$$

To which families of organic chemicals do the compounds belong?

(A) alcohol → aldehyde

(B) alcohol → ketone

(C) ether → ketone

(D) ketone → organic acid

**147.** Which statement INACCURATELY characterizes the health effects of carbon monoxide exposure?

(A) It is lethal at concentrations exceeding 5000 ppm.

(B) It heightens nervousness and mental agitation.

(C) It deprives the body of oxygen.

(D) It forms carboxyhemoglobin with blood.

**148.** Which of the following phrases best describes hazardous air pollutants?

(A) compounds such as particulate matter that decrease visibility

(B) compounds such as hydrocarbons and $NO_x$ that are photochemically oxidized

(C) compounds such as asbestos and benzene that pose a serious risk to human health

(D) compounds such as ozone that contribute to respiratory problems

**149.** What information for chemical handling would NOT be included on a safety data sheet?

(A) spill response measures

(B) disposal of wastes generated from spill response

(C) decontamination of wastes generated from spill response

(D) proper storage procedures

**150.** Which agency has regulatory authority for radon gas exposure in homes, schools, and other buildings?

(A) the Nuclear Regulatory Commission (NRC) under the Nuclear Waste Policy Act of 1982

(B) the Office of Housing and Urban Development (HUD) in association with the U.S. Department of Education under the Indoor Radon Abatement Act of 1988

(C) the U.S. Environmental Protection Agency (EPA) under the Indoor Radon Abatement Act of 1988

(D) none of the above

**151.** What compound is commonly associated with odors from wastewater treatment plants?

(A) $H_2S$ gas produced from biological reduction of sulfates

(B) $CO_2$ gas produced from oxidation of organic matter

(C) $CH_4$ gas produced during anaerobic decomposition of organic matter

(D) $N_2$ produced during denitrification

**152.** The Globally Harmonized System of Classification and Labeling of Chemicals (GHS) uses skull-and-crossbones pictograms to designate acute oral toxicity based on $LD_{50}$. What $LD_{50}$ range is specified for category 2 acute oral toxicity?

(A) $> 5 < 50$

(B) $\geq 50 < 300$

(C) $\geq 300 < 2000$

(D) $\geq 2000 < 5000$

**153.** Strontium-90 has a half-life of 28.8 yr. Most nearly, what is the first order rate constant for strontium-90 decay?

(A) $0.024 \text{ yr}^{-1}$

(B) $0.069 \text{ yr}^{-1}$

(C) $0.26 \text{ yr}^{-1}$

(D) $0.50 \text{ yr}^{-1}$

**154.** Most nearly, what is the time-weighted average noise level for a noise dose of 112%?

(A) 12 dB

(B) 86 dB

(C) 89 dB

(D) 91 dB

**155.** A complete mix activated sludge process, with two parallel secondary clarifiers, will treat wastewater at a total flow rate to the bioreactor of 60 000 m³/d. The influent to the bioreactor has a volatile suspended solids concentration of 100 mg/L. The single bioreactor has a volume 8000 m³, and biomass production in the bioreactor is 0.5 kg/m³·d. The return volatile solids concentration to the bioreactor is 12 000 mg/L. For a mixed liquor volatile suspended solids of 1800 mg/L, what is most nearly the current total return solids flow rate?

(A) 500 m³/d

(B) 4300 m³/d

(C) 9600 m³/d

(D) 12 000 m³/d

**156.** A tire manufacturer generates 11 000 kg/d of shredded vulcanized rubber waste from production operations occurring over two 8 h shifts 5 d/wk. The heating value and ash content, respectively, of the rubber waste is 90 000 kJ/kg and 7%. What is most nearly the hourly heat value of the shredded rubber?

(A) $1.0 \times 10^7$ kJ/h

(B) $2.9 \times 10^7$ kJ/h

(C) $4.1 \times 10^7$ kJ/h

(D) $5.8 \times 10^7$ kJ/h

**157.** The data set shown in the following table represents a 20 yr streamflow record.

| rank | year | flow (ft³/sec) | probability | return period (yr) |
|---|---|---|---|---|
| 1 | 1996 | 3334 | 0.048 | 21.00 |
| 2 | 1997 | 3254 | 0.095 | 10.50 |
| 3 | 1998 | 2608 | 0.143 | 7.00 |
| 4 | 1999 | 2423 | 0.190 | 5.25 |
| 5 | 2000 | 2273 | 0.238 | 4.20 |
| 6 | 2001 | 2123 | 0.286 | 3.5 |
| 7 | 2002 | 2031 | 0.333 | 3.00 |
| 8 | 2003 | 1973 | 0.381 | 2.63 |
| 10 | 2004 | 1904 | 0.476 | 2.10 |
| 10 | 2005 | 1904 | 0.476 | 2.10 |
| 11 | 2006 | 1881 | 0.524 | 1.91 |
| 12 | 2007 | 1558 | 0.571 | 1.75 |
| 13 | 2008 | 1454 | 0.619 | 1.62 |
| 15 | 2009 | 1315 | 0.714 | 1.40 |
| 15 | 2010 | 1315 | 0.714 | 1.40 |
| 16 | 2011 | 1269 | 0.762 | 1.31 |
| 17 | 2012 | 1258 | 0.810 | 1.24 |
| 18 | 2013 | 1223 | 0.857 | 1.17 |
| 19 | 2014 | 1177 | 0.905 | 1.11 |
| 20 | 2015 | 1004 | 0.952 | 1.05 |

What is most nearly the flow that has a 20% risk of being equaled or exceeded at least once in 10 years?

(A) 1180 ft³/sec

(B) 1250 ft³/sec

(C) 1490 ft³/sec

(D) 3100 ft³/sec

**158.** The following table shows a portion of an integrated cost/work schedule. The total project budget is $830,000 for a scheduled duration of 24 days.

| day | schedule used (%) | activity | target activity cost ($/day) | project cost to date ($) | budget spent (%) | work completed (%) |
|---|---|---|---|---|---|---|
| 12 | 50% | D/F | $39,000 | $325,000 | 39% | 35% |
| 13 | 54% | D/F | $39,000 | $365,000 | 43% | 40% |
| 14 | 58% | G | $18,000 | $380,000 | 46% | 45% |

What is the budget and schedule status of the project on day 13?

(A) under budget and ahead of schedule

(B) on budget and behind schedule

(C) over budget and ahead of schedule

(D) over budget and behind schedule

**159.** A project has an initial cost of $1.3M and annual maintenance costs as shown in the following table.

| year | 1 | 2 | 3 |
|---|---|---|---|
| cost ($) | 245,000 | 209,000 | 167,000 |

Capital costs are financed at 2% annual interest, and maintenance costs are financed at 4% annual interest. Most nearly, what is the equivalent uniform annual cost of the project?

(A) $434,000

(B) $513,000

(C) $660,000

(D) $676,000

**160.** Traffic data in metropolitan areas show distinct increases in vehicles on the road during the morning and evening commutes, and this pattern can be described as a bimodal distribution. To visualize this distribution, which statistical plot of vehicles on the road is most appropriate?

(A) box-and-whisker plot

(B) pie chart

(C) histogram

(D) stacked column chart

# STOP!

### DO NOT CONTINUE!

This concludes the Afternoon Session of the examination. If you finish early, check your work and make sure that you have followed all instructions. After checking your answers, submit your solutions and leave the examination room. Once your answers are submitted you will not be able to access them again.

# Answer Keys and Solutions

# Exam 1 Answer Key

## Morning Session 1

| | | | | |
|---|---|---|---|---|
| 1. A | 11. A | 21. B | 31. B |
| 2. B | 12. B | 22. B | 32. A |
| 3. D | 13. C | 23. B | 33. C |
| 4. B | 14. B | 24. D | 34. A |
| 5. A | 15. C | 25. B | 35. A |
| 6. C | 16. B | 26. B | 36. C |
| 7. B | 17. B | 27. D | 37. A |
| 8. A | 18. A | 28. C | 38. D |
| 9. C | 19. B | 29. D | 39. A |
| 10. C | 20. B | 30. B | 40. B |

## Afternoon Session 1

| | | | | |
|---|---|---|---|---|
| 41. D | 51. C | 61. C | 71. C |
| 42. A | 52. B | 62. A | 72. D |
| 43. A | 53. C | 63. B | 73. D |
| 44. A | 54. C | 64. B | 74. B |
| 45. C | 55. B | 65. C | 75. B |
| 46. D | 56. D | 66. B | 76. D |
| 47. B | 57. D | 67. B | 77. D |
| 48. C | 58. D | 68. A | 78. D |
| 49. B | 59. D | 69. D | 79. D |
| 50. A | 60. B | 70. B | 80. A |

# Exam 2 Answer Key

## Morning Session 2

81. C
82. A
83. B
84. B
85. C
86. C
87. B
88. A
89. A
90. C
91. B
92. B
93. B
94. C
95. B
96. B
97. C
98. C
99. C
100. D
101. C
102. A
103. B
104. B
105. A
106. B
107. B
108. B
109. B
110. C
111. A
112. C
113. C
114. B
115. B
116. B
117. A
118. B
119. B
120. B

## Afternoon Session 2

121. D
122. B
123. B
124. C
125. B
126. B
127. D
128. C
129. B
130. B
131. D
132. C
133. C
134. B
135. A
136. B
137. A
138. B
139. D
140. D
141. C
142. B
143. B
144. B
145. D
146. B
147. B
148. C
149. C
150. C
151. A
152. A
153. A
154. D
155. C
156. B
157. A
158. D
159. B
160. C

# Solutions
## Morning Session 1

**1.** During the maximum month, the average daily flow is

$$\frac{\left(9{,}000{,}000 \ \frac{\text{gal}}{\text{yr}}\right)\left(\frac{125\%}{100\%}\right)}{\left(12 \ \frac{\text{mo}}{\text{yr}}\right)\left(30 \ \frac{\text{day}}{\text{mo}}\right)}$$
$$= 31{,}250 \ \text{gal/day} \quad (31{,}000 \ \text{gal/day})$$

**The answer is (A).**

**2.** The required storage volume is

$$V_S = \left(\frac{V_S}{V_R}\right) V_R = (0.25)(1.2 \ \text{cm})$$
$$= 0.30 \ \text{cm}$$

*(handwritten notes: runoff storage ratio; $V_R$ = Runoff volume; $V_S$ = Detention volume)*

**The answer is (B).**

**3.** From *NCEES Handbook* Clarifier, the settling zone surface area per basin is

$$\text{Hydraulic residence time} = V/Q = \theta$$
$$V = Q\theta = AZ_o$$
$$A = \frac{Q\theta}{Z_o(\text{number of basins})}$$
$$= \frac{\left(12\,000 \ \frac{\text{m}^3}{\text{d}}\right)(100 \ \text{min})}{(2.5 \ \text{m})\left(1440 \ \frac{\text{min}}{\text{d}}\right)(4 \ \text{basins})}$$
$$= 83.2 \ \text{m}^2/\text{basin}$$

The overflow rate is

$$v_o = \frac{\dfrac{Q}{\text{basin}}}{A} = \frac{\left(12\,000 \ \frac{\text{m}^3}{\text{d}}\right)\left(\frac{1 \ \text{d}}{24 \ \text{h}}\right)}{\left(83.2 \ \frac{\text{m}^2}{\text{basin}}\right)(4)}$$
$$= 1.5 \ \text{m}^3/\text{m}^2\cdot\text{h}$$

**The answer is (D).**

**4.** The flow rate per tank is

$$\frac{Q}{\text{number of tanks}} = \frac{7500 \ \frac{\text{m}^3}{\text{d}}}{2 \ \text{tanks}}$$
$$= 3750 \ \text{m}^3/\text{d}\cdot\text{tank}$$

From *NCEES Handbook* Clarifier, the weir length is

$$\text{WOR} = Q/\text{Weir Length}$$
$$\text{Weir Length} = \frac{Q}{\text{WOR}}$$
$$= \frac{3750 \ \frac{\text{m}^3}{\text{d}\cdot\text{tank}}}{\left(14 \ \frac{\text{m}^3}{\text{m}\cdot\text{h}}\right)\left(24 \ \frac{\text{h}}{\text{d}}\right)}$$
$$= 11.2 \ \text{m/tank} \quad (11 \ \text{m/tank})$$

**The answer is (B).**

**5.** The BOD:N:P for the pond influent is 200:15:4. Let subscript $c$ represent cells and subscript $o$ represent influent.

The nitrogen requirement is

$$\frac{\text{BOD}_o}{\text{BOD}_c} = \frac{N_o}{N_c}$$
$$\frac{200 \ \frac{\text{mg}}{\text{L}}}{60 \ \frac{\text{mg}}{\text{L}}} = \frac{N_o}{12 \ \frac{\text{mg}}{\text{L}}}$$
$$N_o = 40 \ \frac{\text{mg}}{\text{L}} > 15 \ \frac{\text{mg}}{\text{L}}$$

Therefore, the influent is likely to be deficient in nitrogen.

The phosphorous requirement is

$$\frac{\text{BOD}_o}{\text{BOD}_c} = \frac{P_o}{P_c}$$

$$\frac{200 \frac{\text{mg}}{\text{L}}}{60 \frac{\text{mg}}{\text{L}}} = \frac{P_o}{1 \frac{\text{mg}}{\text{L}}}$$

$$P_o = 3.33 \frac{\text{mg}}{\text{L}} < 4 \frac{\text{mg}}{\text{L}}$$

Therefore, the influent likely contains adequate amounts of phosphorous.

**The answer is (A).**

**6.** From the table "Dissolved Oxygen Concentration in Water" in the *NCEES Handbook*, the saturated dissolved oxygen in the river above the discharge is

$$\text{DO}_{\text{sat}} = 11.55 \text{ mg/L at } 48°\text{F} \ (9°\text{C})$$

From *NCEES Handbook* Stream Modeling, the dissolved oxygen deficit at the discharge is

$$\text{DO} = \text{DO}_{\text{sat}} - D = 11.55 \frac{\text{mg}}{\text{L}} - 8.9 \frac{\text{mg}}{\text{L}}$$
$$= 2.65 \text{ mg/L}$$

From *NCEES Handbook* Stream Modeling, the critical time is

$$t_c = \frac{1}{(k_2 - k_1)} \ln\left[\left(\frac{k_2}{k_1}\right)\left(1 - \frac{D_o(k_2 - k_1)}{k_1 L_0}\right)\right]$$

$$= \left(\frac{0.23}{\text{d}} - \frac{0.16}{\text{d}}\right)^{-1}$$

$$\times \ln\left(\left(\frac{\frac{0.23}{\text{d}}}{\frac{0.16}{\text{d}}}\right)\left(1 - \frac{\left(2.65 \frac{\text{mg}}{\text{L}}\right) \times \left(\frac{0.23}{\text{d}} - \frac{0.16}{\text{d}}\right)}{\left(\frac{0.16}{\text{d}}\right)\left(10 \frac{\text{mg}}{\text{L}}\right)}\right)\right)$$

$$= 3.40 \text{ d} \quad (3.4 \text{ d})$$

**The answer is (C).**

**7.** From the table "CT Values for 3-LOG Inactivation of Giardia Cysts by Free Chlorine" in the *NCEES Handbook*, for 2.0 mg/L free chlorine concentration at 15°C and 6.5 pH, CT is 69 min·mg/L. From the equation for CT values in *NCEES Handbook* Secondary Drinking Water Standards, the minimum hydraulic resident time is

$$CT = \text{concentration} \times \text{time}$$

$$\text{time} = \frac{CT}{\text{concentration}}$$

$$= \frac{69 \frac{\text{min·mg}}{\text{L}}}{2.0 \frac{\text{mg}}{\text{L}}}$$

$$= 34.5 \text{ min} \quad (35 \text{ min})$$

**The answer is (B).**

**8.** The solids residence time for design is

$$\theta_c = \theta_{c,\min}(\text{SF}) = (3.0 \text{ d})(2.5)$$
$$= 7.5 \text{ d}$$

The biosolids production is

$$Y_{\text{obs}}(S_o - S)Q = (0.36)\left(192 \frac{\text{mg}}{\text{L}} - 20 \frac{\text{mg}}{\text{L}}\right)$$
$$\times \left(5000 \frac{\text{m}^3}{\text{d}}\right)\left(10^{-3} \frac{\text{kg·L}}{\text{mg·m}^3}\right)$$
$$= 310 \text{ kg/d}$$

**The answer is (A).**

**9.** From *NCEES Handbook* Circular Pipe Head Loss Equation (Head Loss Expressed in Feet),

$$h_f = \frac{4.73L}{C^{1.852}D^{4.87}}Q^{1.852}$$

Expressed in terms of the pipe diameter,

$$D = \left(\frac{4.73L}{C^{1.852}h_f}Q^{1.852}\right)^{1/4.87}$$

From the table "Values of Hazen-Williams Coefficient C" in the *NCEES Handbook*, C = 120.

$$D = \left(\left(\frac{(4.73)(1373 \text{ ft})}{(120)^{1.852}(1.2 \text{ ft})}\right)\left(0.5 \frac{\text{ft}^3}{\text{sec}}\right)^{1.852}\right)^{1/4.87}$$

$$= (0.73 \text{ ft})\left(12 \frac{\text{in}}{\text{ft}}\right)$$

$$= 8.7 \text{ in}$$

The nearest standard pipe size larger than 8.7 in is 10 in.

**The answer is (C).**

**10.** Let $t_m$ be the time corresponding to the peak of the concentration time plot.

From the illustration, $t_m = 40$ min.

*The answer is (C).*

**11.** The average annual irrigation water demand is

$$(0.20)(20 \text{ ac})\left(26 \frac{\text{wk}}{\text{yr}}\right)\left(1 \frac{\text{in}}{\text{wk}}\right)\left(\frac{1 \text{ ft}}{12 \text{ in}}\right)$$
$$\times \left(43{,}560 \frac{\text{ft}^2}{\text{ac}}\right)\left(7.48 \frac{\text{gal}}{\text{ft}^3}\right)$$
$$= 2{,}823{,}850 \text{ gal/yr} \quad (2{,}800{,}000 \text{ gal/yr})$$

*The answer is (A).*

**12.**

| use | people | gal/day | days/wk | gal/wk |
|---|---|---|---|---|
| apartment | 200 | 100 | 7 | 140,000 |
| office | 100 | 15 | 5 | 7500 |
| restaurant | (2)(62) | 9 | 7 | 7812 |
| deli | 36 | 6 | 5 | 1080 |
| club | (0.4)(200) | 100 | 7 | 56,000 |
|  | (0.4)(100) | 100 | 5 | 20,000 |
|  |  |  |  | 232,392 |

$$\left(232{,}392 \frac{\text{gal}}{\text{wk}}\right)\left(52 \frac{\text{wk}}{\text{yr}}\right) = 12{,}084{,}384 \text{ gal/yr}$$
$$(12{,}000{,}000 \text{ gal/yr})$$

*The answer is (B).*

**13.** From *NCEES Handbook* Rapid Mix and Flocculator Design,

$$G = \sqrt{\frac{P}{\mu V}}$$
$$P = G^2 \mu V$$
$$= \left(\frac{45}{\text{s}}\right)^2 (100 \text{ m}^3)\left(0.001002 \frac{\text{N}\cdot\text{s}}{\text{m}^2}\right)$$
$$\times \left(\frac{1 \text{ kW}\cdot\text{s}}{1000 \text{ N}\cdot\text{m}}\right)$$
$$= 0.203 \text{ kW} \quad (0.20 \text{ kW})$$

*The answer is (C).*

**14.** See *NCEES Handbook* BOD Test Solution and Seeding Procedures. Disregard sample 1 because $DO_5$ is below 2.0 mg/L (possible anaerobic activity), and disregard sample 4 because $DO_5$ is above 7.0 mg/L (insufficient aerobic activity).

For sample 2,

$$\text{BOD, mg/L} = \frac{D_1 - D_2}{P}$$
$$= \frac{9.2 \frac{\text{mg}}{\text{L}} - 2.3 \frac{\text{mg}}{\text{L}}}{\frac{100 \text{ mL}}{300 \text{ mL}}}$$
$$= 20.7 \text{ mg/L}$$

For sample 3,

$$\text{BOD, mg/L} = \frac{9.3 \frac{\text{mg}}{\text{L}} + 5.8 \frac{\text{mg}}{\text{L}}}{\frac{50 \text{ ml}}{300 \text{ ml}}} = 21.0 \text{ mg/L}$$

$BOD_5$ at 25°C is

$$\text{BOD, mg/L} = \frac{20.7 \frac{\text{mg}}{\text{L}} + 21.0 \frac{\text{mg}}{\text{L}}}{2} = 20.85 \text{ mg/L}$$

From *NCEES Handbook* Microbial Kinetics: BOD Exertion,

$$y_t = L(1 - e^{-kt})$$
$$L = \frac{y_t}{1 - e^{-kt}}$$
$$= \frac{20.85 \frac{\text{mg}}{\text{L}}}{1 - e^{-(0.4/\text{d})(5 \text{ d})}}$$
$$= 24.1 \text{ mg/L}$$

From *NCEES Handbook* Microbial Kinetics: Kinetic Temperature Corrections,

$$k_T = k_{25} \theta^{(T-25)}$$
$$k_{20} = \left(\frac{0.40}{\text{d}}\right)(1.047)^{20-25}$$
$$= 0.32/\text{d}$$

$BOD_5$ at 20°C is

$$\left(24.1 \frac{\text{mg}}{\text{L}}\right)(1 - e^{-(0.32/\text{d})(5\text{d})}) = 19.2 \text{ mg/L}$$
$$(19 \text{ mg/L})$$

*The answer is (B).*

**15.** From the table "Dissolved Oxygen Concentration in Water" from the *NCEES Handbook*,

$$DO_{sat} = 11.28 \text{ mg/L at } 50°F \ (10°C)$$

From the DO sag curve, and using the equation for dissolved oxygen concentration from *NCEES Handbook* Microbial Kinetics: Stream Modeling, the critical DO deficit occurs at about 3.5 days and is equal to

$$DO = DO_{sat} - D$$
$$= 11.28 \frac{\text{mg}}{\text{L}} - 6.1 \frac{\text{mg}}{\text{L}}$$
$$= 5.18 \text{ mg/L} \quad (5.2 \text{ mg/L})$$

**The answer is (C).**

**16.** The GAC change-out interval is

$$\frac{20{,}000 \dfrac{\text{lbm GAC}}{\text{vessel}}}{500 \dfrac{\text{lbm GAC}}{\text{day}}} = 40 \text{ days/vessel}$$

**The answer is (C).**

**17.** The ratio of post-development to pre-development flow is 1.5:1 or 150%. For a relatively small drainage basin of 100 ha, increasing runoff by 150%, or by 0.5 m³/s, will likely considerably increase flow and justify stormwater control and containment devices.

**The answer is (B).**

**18.** In pond 2,

$$\left(2000 \frac{\text{mg}}{\text{L}}\right)(0.50)(0.80) = 800 \text{ mg/L BOD removal}$$

$$\left(800 \frac{\text{mg}}{\text{L}}\right)\left(400 \frac{\text{m}^3}{\text{d}}\right)\left(10^{-6} \frac{\text{kg}}{\text{mg}}\right)$$
$$\times \left(10^3 \frac{\text{L}}{\text{m}^3}\right)\left(\frac{1 \text{ d}}{24 \text{ h}}\right)$$
$$= 13.3 \text{ kg BOD/h}$$
$$= 13.3 \text{ kg DO/h} \quad (13 \text{ kg DO/h})$$

$$\begin{bmatrix}\text{assuming a 1:1 ratio of dissolved} \\ \text{oxygen required for BOD removal}\end{bmatrix}$$

**The answer is (A).**

**19.** From the illustration at $Z_o = 2.5$ m and $t = 100$ min, $h_o = 64\%$. At $Z_o = 2.0$ m and $t = 100$ min, $h_o = 69\%$.

The efficiency increases.

**The answer is (B).**

**20.** The initial pH is well below the 8.3 pH system point. Therefore, essentially all alkalinity is present as bicarbonate ($HCO_3^-$) as illustrated.

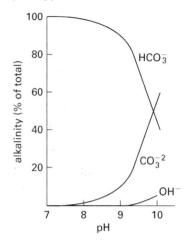

**The answer is (B).**

**21.** From the tables "Common Radicals in Water" and "Periodic Table of Elements" from the *NCEES Handbook*,

| ion | mg/L | mg/mmol | meq/mmol | meq/L |
|---|---|---|---|---|
| $SO_4^{-2}$ | 17 | 96 | 2 | 0.35 |
| $Na^+$ | 48 | 23 | 1 | 2.1 |
| $K^+$ | 14 | 39 | 1 | 0.36 |
| $Cl^-$ | 39 | 35 | 1 | 1.1 |
| $Ca^{+2}$ | 56 | 40 | 2 | 2.8 |
| $Mg^{+2}$ | 12 | 24 | 2 | 1.0 |
| $HCO_3^-$ | 237 | 61 | 1 | 3.9 |

The sum of the anions is

$$0.35 \frac{\text{meq}}{\text{L}} + 1.1 \frac{\text{meq}}{\text{L}} + 3.9 \frac{\text{meq}}{\text{L}} = 5.35 \text{ meq/L}$$

The sum of the cations is

$$2.1 \frac{\text{meq}}{\text{L}} + 0.36 \frac{\text{meq}}{\text{L}} + 2.8 \frac{\text{meq}}{\text{L}} + 1.0 \frac{\text{meq}}{\text{L}} = 6.26 \text{ meq/L}$$

The percent difference is

$$\left(\frac{\sum \text{anions} - \sum \text{cations}}{\sum \text{anions} + \sum \text{cations}}\right) \times 100\%$$
$$= \left(\frac{5.35 \dfrac{\text{meq}}{\text{L}} - 6.26 \dfrac{\text{meq}}{\text{L}}}{5.35 \dfrac{\text{meq}}{\text{L}} + 6.26 \dfrac{\text{meq}}{\text{L}}}\right) \times 100\%$$
$$= 7.84\%$$

Because 7.84% > 2%, for range of the sum of the anions between 3.0 meq/L and 10.0 meq/L, the analysis is not complete. It is deficient in anions.

**The answer is (B).**

**22.** The ratio of specific conductivity to total dissolved solids is about 0.075. This ratio is typically about 1.7. Because the specific conductance is much less than it would be if it were composed of ionic solids, the majority of the solids in the sample are likely nonionic.

**The answer is (B).**

**23.** From *NCEES Handbook* National Research Council (NRC) Trickling Filter Performance,

$$E_1 = \frac{100}{1 + 0.0561\sqrt{\frac{W}{VF}}}$$

Find the BOD efficiency.

$$E_{BOD} = \frac{186\,\frac{mg}{L} - 30\,\frac{mg}{L}}{186\,\frac{mg}{L}} \times 100\%$$
$$= 84\%$$

Find the BOD loading.

$$W = \left(186\,\frac{mg}{L}\right)(0.25\text{ MGD})\left(10^6\,\frac{gal}{day}\right)$$
$$\times \left(3.785\,\frac{L}{gal}\right)\left(0.454\,\frac{lbm}{10^6\,mg}\right)$$
$$= 80\text{ lbm/day}$$

Substitute the values into the NRC equation and solve for filter volume, $V$.

$$84\% = \frac{100}{1 + 0.0561\sqrt{\frac{80\,\frac{lbm}{day}}{V2}}}$$

$$V = (13.9)(10^3\text{ ft}^3)$$
$$= 13{,}900\text{ ft}^3 \quad (14{,}000\text{ ft}^3)$$

**The answer is (B).**

**24.** From *NCEES Handbook* Methanol Requirement for Biologically Treated Wastewater, the required methanol concentration is

$$C_m = 2.47N_o + 1.53N_1 + 0.87D_o$$
$$= (2.47)\left(42\,\frac{mg}{L}\right) + (1.53)\left(13\,\frac{mg}{L}\right) + (0.87)\left(1.9\,\frac{mg}{L}\right)$$
$$= 125\text{ mg/L}$$

Calculating the daily volume, $V$, of methanol gives

$$\left(2.5 \times 10^6\,\frac{gal}{day}\right)\left(125\,\frac{mg}{L}\right)\left(3.785\,\frac{L}{gal}\right)\left(10^6\,\frac{mg}{kg}\right)$$
$$\times \left(1\,\frac{L}{kg}\right)\left(\frac{1}{0.7915}\right)$$
$$= 1490\text{ L/d}$$

**The answer is (D).**

**25.** From the equation for partial pressure from *NCEES Handbook* Ideal Gas Mixtures: Partial Pressures, if $P = \sum P_i$, $\sum P_{i,\%} = 100\%$

$$P = \sum P_i$$
$$\sum \text{percentage of each gas} = 100\%$$

$$\frac{P_{methane}}{P_{total}} = \frac{\text{percentage of methane in gas}}{100\%}$$

$$P_{methane} = P_{total}\left(\frac{\text{percentage of methane in gas}}{100\%}\right)$$
$$= (0.98\text{ atm})\left(\frac{12\%}{100\%}\right)$$
$$= 0.118\text{ atm}$$

**The answer is (B).**

**26.** For each pollutant included in the National Ambient Air Quality Standards (NAAQS), primary and secondary standards are defined. Primary standards are intended to protect human health, and secondary standards are intended to protect public welfare.

Air pollutants may be classified as primary and secondary. Primary pollutants are those that exist in the air in the same form in which they were emitted. Secondary pollutants are those that are formed in the air from other emitted compounds. Primary and secondary pollutants are not the same as, and should not be confused with, the primary and secondary NAAQS.

**The answer is (B).**

**27.** The mass removed is

$$\frac{C_p Q_g E\%}{100\%} = \frac{\left(20 \frac{g}{m^3}\right)\left(14 \frac{m^3}{s}\right)(80\%)}{(100\%)\left(\frac{1 \text{ d}}{86\,400 \text{ s}}\right)\left(1000 \frac{g}{kg}\right)}$$
$$= 19\,354 \text{ kg/d} \quad (19\,000 \text{ kg/d})$$

**The answer is (D).**

**28.** From *NCEES Handbook* Other Than Standard Air, the dry air density is

$$d = 1.325 \frac{P_b}{T}$$
$$= (1.325)\left(\frac{29.81 \text{ in Hg}}{58°F + 460°}\right)$$
$$= 0.076 \text{ lbf/ft}^3$$

**The answer is (C).**

**29.** The sulfur feed to the plant is

$$\frac{\left(8.3 \frac{\text{lbm}}{\text{sec}}\right)(2.7\%)\left(86{,}400 \frac{\text{sec}}{\text{day}}\right)}{100\%} = 19{,}362 \text{ lbm/day}$$

The sulfur emitted to APC equipment is

$$\frac{\left(19{,}362 \frac{\text{lbm}}{\text{day}}\right)(100\% - 5\%)}{100\%} = 18{,}394 \frac{\text{lbm}}{\text{day}}$$

$$S + O_2 \rightarrow SO_2$$

From *NCEES Handbook* table "Periodic Table of Elements,"

$$S \text{ MW} = 32 \text{ g/mol}$$
$$SO_2 \text{ MW} = 32 \frac{g}{\text{mol}} + (2)\left(16 \frac{g}{\text{mol}}\right)$$
$$= 64 \text{ g/mol}$$

$$SO_2 \text{ emitted} = \frac{\left(18{,}394 \frac{\text{lbm}}{\text{day}} \text{ as S}\right)\left(64 \frac{g}{\text{mol}} \text{ as } SO_2\right)}{32 \frac{g}{\text{mol}} \text{ as S}}$$
$$= 36{,}788 \text{ lbm/day} \quad (37{,}000 \text{ lbm/day})$$

**The answer is (D).**

**30.** Most nitrogen oxide (NO) emissions occur from fuel release and fossil fuel combustion as relatively benign NO. However, NO can rapidly oxidize to nitrogen dioxide ($NO_2$), which is associated with respiratory ailments and which reacts with hydrocarbons in the presence of sunlight to form photochemical oxidants. $NO_2$ also reacts with the hydroxyl radical (•OH) in the atmosphere to form nitric acid ($HNO_3$), a contributor to acid rain.

When combusted, fossil fuels, mostly coal, release sulfur primarily as sulfur dioxide ($SO_2$) with much smaller amounts of sulfur trioxide ($SO_3$). Sulfur oxide aerosols may contribute to particulate matter concentrations in the air and are a constituent of acid rain. Sulfur oxides can significantly impact visibility. Sulfur dioxide also reacts with oxygen to form sulfur trioxide ($SO_3$), which subsequently reacts with water to form sulfuric acid ($H_2SO_4$). Sulfur dioxide may react directly with water in the atmosphere to form sulfurous acid ($H_2SO_3$).

**The answer is (B).**

**31.** The pressure drops from the nozzle inlet value to zero as a free jet at the nozzle outlet.

$$-\Delta P = 0.05 \text{ atm} - 0 \text{ atm} = 0.05 \text{ atm}$$

**The answer is (B).**

**32.** The illustration shows an inversion to 500 m and subadiabatic conditions above 500 m.

**The answer is (A).**

**33.** From the table "Atmospheric Stability Under Various Conditions" in the *NCEES Handbook*, for a wind speed of 2.6 m/s measured 10 m above ground level and moderate incoming solar radiation, the stability category is B.

The wind speed is $u = 2.6$ m/s, and the emission rate is $Q = 5.4$ kg/s.

From the figure "Downwind distance where the maximum concentration occurs" in the *NCEES Handbook*, for stability category B and a stack height of 150 m,

$$(Cu/Q)_{max} = 8 \times 10^{-6} \frac{1}{m^2}$$

$$C = \frac{\left(8 \times 10^{-6} \frac{1}{m^2}\right)Q}{u}$$

$$= \frac{\left(8 \times 10^{-6} \frac{1}{m^2}\right)\left(5.4 \frac{\text{kg}}{\text{s}}\right)\left(10^6 \frac{\text{mg}}{\text{kg}}\right)}{2.6 \frac{m}{s}}$$

$$= 16.6 \text{ mg/m}^3 \quad (17 \text{ mg/m}^3)$$

**The answer is (C).**

**34.** The bubble policy allows industry flexibility by treating all activities in a single plant or among a group of proximate industries to emit at various rates as long as the resulting total emissions do not exceed the allowable emissions for each individual source.

*The answer is (B).*

**35.** From *NCEES Handbook* Fabric Filtration, the depth of the dust layer is

$$D_p = \frac{LVt}{\rho_L}$$

$$= \frac{\left(25 \frac{\text{g}}{\text{m}^3}\right)\left(3.5 \frac{\text{m}}{\text{min}}\right)(60 \text{ min})\left(100 \frac{\text{cm}}{\text{m}}\right)}{1.2 \times 10^6 \frac{\text{g}}{\text{m}^3}}$$

$$= 0.44 \text{ cm}$$

*The answer is (A).*

**36.**

$$(8 \text{ compartments})\left(\frac{60 \text{ bags}}{1 \text{ compartment}}\right)\left(\frac{10 \text{ m}^2}{1 \text{ bag}}\right)$$

$$= 4800 \text{ m}^2$$

*The answer is (C).*

**37.** Electrostatic precipitators are used to remove particulate matter from gas streams. They are popular because they are economical to operate, provide high removal efficiencies (near 99%), are dependable and predictable, and do not produce a moisture plume. Electrostatic precipitators generally cannot be used with moist flows and mists, since performance is inhibited by water droplets that can insulate particles and reduce their resistivities.

*The answer is (B).*

**38.** From *NCEES Handbook* Selected Properties of Air, the dry adiabatic lapse rate is $0.0098°C/m$.

The air temperature at 93 m is

$$19°C + \left(-0.0047° \frac{C}{m}\right)(93 \text{ m}) = 18.6°C$$

$$31°C + \left(-0.0098° \frac{C}{m}\right)z = 18.6°C + \left(-0.0047° \frac{C}{m}\right)z$$

The height above the stack where plume and air temperatures are equal is

$$z = \frac{31°C - 18.6°C}{0.0098° \frac{C}{m} - 0.0047° \frac{C}{m}} = 2431 \text{ m}$$

The plume will rise to

$$2431 \text{ m} + 93 \text{ m} = 2524 \text{ m} \quad (2520 \text{ m})$$

*The answer is (D).*

**39.** Radon gas is a naturally occurring gas that can cause lung cancer.

*The answer is (A).*

**40.** The plume moves with the surrounding wind at 2.1 m/s. The center of the plume will arrive at the house in approximately

$$\frac{(1.4 \text{ km})\left(1000 \frac{\text{m}}{\text{km}}\right)}{\left(2.1 \frac{\text{m}}{\text{s}}\right)\left(60 \frac{\text{s}}{\text{min}}\right)} = 11 \text{ min}$$

The leading edge of the plume will arrive before 11 min, and the residents will have to evacuate before then to avoid exposure. Therefore, the time for evacuation is less than 11 min.

*The answer is (B).*

# Solutions
## Afternoon Session 1

**41.** From *NCEES Handbook* Heats of Reaction, the combustion reaction is

$$\Delta H°_{reaction} = \Delta H°_{products} - \Delta H°_{reactants}$$
$$= (1 \text{ mol})\left(-393.5 \frac{\text{kJ}}{\text{mol}}\right)$$
$$+ (2 \text{ mol})\left(-241.8 \frac{\text{kJ}}{\text{mol}}\right)$$
$$- (1 \text{ mol})\left(-74.82 \frac{\text{kJ}}{\text{mol}}\right)$$
$$- (2 \text{ mol})\left(0 \frac{\text{kJ}}{\text{mol}}\right)$$
$$= -802.3 \frac{\text{kJ}}{\text{mol}}$$

The reaction is exothermic.

The heating value per 1000 ft³ of CH₄ at 20°C and 1 atm is

$$\frac{(1000 \text{ ft}^3)\left(28.3 \frac{\text{L}}{\text{ft}^3}\right)\left(802.3 \frac{\text{kJ}}{\text{mol}}\right)}{22.03 \frac{\text{L}}{\text{mol}}}$$
$$= 1.0 \times 10^6 \text{ kJ}/1000 \text{ ft}^3$$

**The answer is (D).**

**42.** Controlled air incinerators are relatively low-technology devices that have difficulty meeting air emission limits.

**The answer is (A).**

**43.** Assume a 100 kg sample.

| waste component | discarded mass (kg) | unit discarded energy ($10^6$ kJ/100 kg) | ash mass (kg/100 kg) |
|---|---|---|---|
| food | 13 | 0.060 | 0.65 |
| glass | 6 | 0.00090 | 5.9 |
| plastic | 4 | 0.13 | 0.4 |
| paper | 37 | 0.62 | 2.2 |
| cardboard | 10 | 0.16 | 0.50 |
| textiles | 1 | 0.017 | 0.025 |
| ferrous metal | 8 | 0.0056 | 7.8 |
| non-ferrous metal | 2 | 0.0014 | 1.9 |
| wood | 4 | 0.074 | 0.060 |
| yard clippings | 15 | 0.098 | 0.68 |
|  | 100 | 1.17 | 20 |

$$\text{discarded mass, kg} = (100 \text{ kg})\left(\frac{\% \text{ mass}}{100}\right)$$

$$\begin{aligned}\text{unit discarded energy,} \\ 10^6 \text{ kJ}/100 \text{ kg}\end{aligned} = (\text{discarded mass, kg})$$
$$\times \begin{pmatrix}\text{component discarded} \\ \text{energy, kJ/kg}\end{pmatrix}$$

$$\text{ash mass, kg}/100 \text{ kg} = (\text{discarded mass, kg})$$
$$\times \left(\frac{\% \text{ ash}}{100}\right)$$

$$\left(\frac{1.17 \times 10^6 \text{ kJ}}{100 \text{ kg discarded mass}}\right)\left(\frac{100 \text{ kg discarded mass}}{80 \text{ kg discarded mass without ash}}\right)$$
$$\times \left(\frac{(10)(100 \text{ kg})}{1000 \text{ kg}}\right)$$
$$= 14.6 \times 10^6 \text{ kJ}/1000 \text{ kg} \quad (1.5 \times 10^7 \text{ kJ}/1000 \text{ kg})$$

**The answer is (A).**

**44.** The compacted waste without cover is

$$\frac{(13\,000 \text{ people})\left(1.2 \frac{\text{kg}}{\text{person}\cdot\text{d}}\right)}{1100 \frac{\text{kg}}{\text{m}^3}} = 14.2 \text{ m}^3/\text{d}$$

Using 1 m³ of cover for every 5 m³ of waste, the compacted waste with cover is

$$\frac{14.2 \frac{\text{m}^3}{\text{d}}}{5} + 14.2 \frac{\text{m}^3}{\text{d}} = 17 \text{ m}^3/\text{d}$$

Assume 365 d/yr of operation.

$$\left(17 \frac{\text{m}^3}{\text{d}}\right)\left(365 \frac{\text{d}}{\text{yr}}\right)(25 \text{ yr}) = 160 \times 10^3 \text{ m}^3$$

**The answer is (A).**

**45.** The minimum required bed incinerator area is

$$\frac{(\text{HV})(\text{fueling rate})}{\text{heat release rate}} = \frac{\left(500 \frac{\text{kg}}{\text{h}}\right)\left(57\,000 \frac{\text{kJ}}{\text{kg}}\right)}{2.3 \times 10^6 \frac{\text{kJ}}{\text{m}^2\cdot\text{h}}}$$

$$= 12.4 \text{ m}^2 \quad (12 \text{ m}^2)$$

**The answer is (C).**

**46.** From *NCEES Handbook* Boiling,

$$q'' = h\Delta T_e$$
$$h = mc$$

| | | |
|---|---|---|
| $c$ | specific heat of water | kJ/kg·°C |
| $m$ | mass of water | kg |
| $q''$ | heat required to raise temperature to 100°C | kJ |
| $q$ | heat required to vaporize water at its boiling point (heat of vaporization) | kJ/kg |

$$c = 4.184 \text{ kJ/kg·°C}$$
$$q = 2258 \text{ kJ/kg}$$

The temperature change is

$$\Delta T_e = 100°C - 20°C$$
$$= 80°C$$
$$m = \left(4 \frac{\text{m}^3}{\text{d}}\right)\left(1000 \frac{\text{kg}}{\text{m}^3}\right)(0.76)\left(\frac{1 \text{ d}}{24 \text{ h}}\right)$$
$$= 127 \text{ kg/h}$$
$$q'' = \left(4.184 \frac{\text{kJ}}{\text{kg·°C}}\right)\left(127 \frac{\text{kg}}{\text{h}}\right)(80°C)$$
$$= 42\,509 \text{ kJ/h}$$

The total heat to dry the sludge is

$$q_T = q'' + q$$
$$= 42\,509 \frac{\text{kJ}}{\text{h}} + \left(2258 \frac{\text{kJ}}{\text{kg}}\right)\left(127 \frac{\text{kg}}{\text{h}}\right)$$
$$= 329\,275 \text{ kJ/h} \quad (330\,000 \text{ kJ/h})$$

**The answer is (D).**

**47.** From the table "Periodic Table of Elements" in the *NCEES Handbook*,

$$\text{Ca}^{+2} \text{ MW} = 40 \text{ mg/mmol}$$

$$\frac{187 \frac{\text{mg}}{\text{L}}}{40 \frac{\text{mg}}{\text{mmol}}} = 4.675 \text{ mmol/L}$$

From the reaction equation, 4.675 mmol/L $\text{Ca}^{+2}$ reacts to produce 4.675 mmol/L $\text{CaCO}_3$.

$$\text{CaCO}_3 \text{ MW} = 40 \frac{\text{g}}{\text{mol}} + 12 \frac{\text{g}}{\text{mol}} + (3)\left(16 \frac{\text{g}}{\text{mol}}\right)$$
$$= 100 \text{ g/mol} \quad (100 \text{ mg/mmol})$$

$$\left(100 \frac{\text{mg}}{\text{mmol}}\right)\left(4.675 \frac{\text{mmol}}{\text{L}}\right)\left(0.7 \frac{\text{m}^3}{\text{min}}\right)\left(1000 \frac{\text{L}}{\text{m}^3}\right)$$
$$\times \left(\frac{1 \text{ kg}}{10^6 \text{ mg}}\right)\left(960 \frac{\text{min}}{\text{d}}\right)\left(240 \frac{\text{d}}{\text{yr}}\right)$$
$$= 75\,398 \text{ kg/yr} \quad (75\,000 \text{ kg/yr})$$

**The answer is (B).**

**48.** See *NCEES Handbook* Acids, Bases, and pH. At pH 8.6,

$$[\text{OH}^-] = 10^{-(14-8.6)} \frac{\text{mol}}{\text{L}} = 3.98 \times 10^{-6} \text{ mol/L}$$

Assume a waste temperature of 25°C. From *NCEES Handbook* Chemistry Definitions, at 25°C, the solubility product constant is

$$K_{SP} = [\text{Cd}^{+2}][\text{OH}^-]^2 = 5.27 \times 10^{-15}$$

The $\text{Cd}^{+2}$ concentration at pH 8.6 is

$$[\text{Cd}^{+2}] = \frac{5.27 \times 10^{-15}}{\left(3.98 \times 10^{-6} \frac{\text{mol}}{\text{L}}\right)^2}$$
$$= 3.33 \times 10^{-4} \text{ mol/L}$$

$$\left(3.33 \times 10^{-4} \frac{\text{mol}}{\text{L}}\right)\left(112.4 \frac{\text{g}}{\text{mol}}\right)\left(1000 \frac{\text{mg}}{\text{g}}\right)$$
$$= 37.4 \text{ mg/L} \quad (37 \text{ mg/L})$$

**The answer is (C).**

**49.** The total daily mass of waste generated by the community on an as-discarded basis is

$$(37{,}000 \text{ people})\left(2.2 \frac{\text{lbm}}{\text{person-day}}\right) = 81{,}400 \text{ lbm/day}$$

During a 5 day work week, the number of loads collected by a single truck is

$$\left(3 \frac{\text{loads}}{\text{truck-day}}\right)\left(5 \frac{\text{day}}{\text{wk}}\right) = 15 \text{ loads/truck-wk}$$

$$\frac{\left(81{,}400 \frac{\text{lbm}}{\text{day}}\right)\left(7 \frac{\text{day}}{\text{wk}}\right)}{\left(760 \frac{\text{lbm}}{\text{yd}^3}\right)\left(12 \frac{\text{yd}^3}{\text{load}}\right)} = 62.48 \text{ loads/wk} \quad (63 \text{ loads/wk})$$

The required number of trucks is

$$\frac{63 \frac{\text{loads}}{\text{wk}}}{15 \frac{\text{loads}}{\text{truck-wk}}} = 4.2 \text{ trucks} \quad (5 \text{ trucks})$$

**The answer is (B).**

**50.** From the table "Common Radicals in Water" in the *NCEES Handbook*,

$$\text{CO}_2 \text{ MW} = 44 \text{ g/mol}$$

$$(0.25)\left(60{,}000 \frac{\text{ft}^3}{\text{day}}\right)\left(\frac{1 \text{ mol}}{24.4 \text{ L}}\right)\left(28.2 \frac{\text{L}}{\text{ft}^3}\right)$$
$$= 17{,}336 \text{ mol/d}$$

$$m = \left(17{,}336 \frac{\text{mol}}{\text{d}}\right)\left(44 \frac{\text{g}}{\text{mol}}\right)\left(\frac{1 \text{ kg}}{1000 \text{ g}}\right)$$
$$= 763 \text{ kg/d} \quad (760 \text{ kg/d})$$

**The answer is (A).**

**51.** The hydraulic loading rate can be increased by increasing the flow to the exchangers or by decreasing the exchanger bed area. Bed area is typically fixed by standard vessel sizing, but flow can be increased by recirculation.

**The answer is (C).**

**52.** From *NCEES Handbook* Break-Through Time for Leachate to Penetrate a Landfill Clay Liner, the breakthrough time is

$$t = \frac{d^2\eta}{K(d+h)}$$
$$= \frac{(3 \text{ ft})^2(0.15)}{\left(0.084 \frac{\text{ft}}{\text{yr}}\right)(3 \text{ ft} + 2.3 \text{ ft})}$$
$$= 3.03 \text{ yr} \quad (3 \text{ yr})$$

**The answer is (B).**

**53.** From *NCEES Handbook* table "Periodic Table of the Elements,"

$$\text{CrO}_3 \text{ MW} = 52 \frac{\text{g}}{\text{mol}} + (3)\left(16 \frac{\text{g}}{\text{mol}}\right)$$
$$= 100 \text{ g/mol} \quad (100 \text{ mg/mmol})$$

$$\frac{534 \frac{\text{mg}}{\text{L}}}{100 \frac{\text{mg}}{\text{mmol}}} = 5.34 \frac{\text{mmol}}{\text{L}}$$

$$\text{NaHSO}_3 \text{ MW} = 23 \frac{\text{g}}{\text{mol}} + 1 \frac{\text{g}}{\text{mol}} + 32 \frac{\text{g}}{\text{mol}}$$
$$+ (3)\left(16 \frac{\text{g}}{\text{mol}}\right)$$
$$= 104 \text{ g/mol} \quad (104 \text{ mg/mmol})$$

From the reduction reaction, $(4/4)(5.34 \text{ mmol/L})$ of $\text{CrO}_3$ will react with $(6/4)(5.34 \text{ mmol/L})$ of $\text{NaHSO}_3$.

$$\left(\frac{6}{4}\right)\left(5.34 \frac{\text{mmol}}{\text{L}}\right) = 8.01 \text{ mmol/L}$$

The facility operates 24 h/d for 365 d/yr.

$$\left(104 \ \frac{\text{mg}}{\text{mmol}}\right)\left(8.01 \ \frac{\text{mmol}}{\text{L}}\right)\left(1.8 \ \frac{\text{m}^3}{\text{min}}\right)$$
$$\times \left(10^{-3} \ \frac{\text{kg} \cdot \text{L}}{\text{mm} \cdot \text{m}^3}\right)\left(1400 \ \frac{\text{min}}{\text{d}}\right)\left(365 \ \frac{\text{d}}{\text{yr}}\right)$$
$$= 7.88 \times 10^5 \ \text{kg/yr} \quad [\text{for 100\% pure NaHSO}_3]$$

$$\frac{\left(7.88 \times 10^5 \ \frac{\text{kg}}{\text{yr}}\right)\left(\frac{\$120}{1000 \ \text{kg}}\right)}{0.92}$$
$$= \$102{,}870/\text{yr} \quad (\$100{,}000/\text{yr})$$

**The answer is (C).**

**54.** Per *NCEES Handbook* Carcinogens, for carcinogens, it is assumed that any dose of a carcinogen will result in a risk. Per *NCEES Handbook* Noncarcinogens, for noncarcinogens, a threshold dose is assumed, below which no risk occurs.

**The answer is (C).**

**55.** From the table "Frequency Multiplier Table" in the *NCEES Handbook*, for LF = 0.5/min at up to 8 hr/day, FM = 0.81.

From the table "Coupling Multiplier Table" in the *NCEES Handbook*, for an optimally designed container with no cut-out handles and V ≥ 36 in, CM = 1.00.

From *NCEES Handbook* Recommended Weight Limit, RWL,

$$\text{RWL} = 51(10/\text{H})(1 - 0.0075\,|\text{V}{-}30|)$$
$$\times (0.82 + 1.8/\text{D})(1 - 0.0032\text{A})(\text{FM})(\text{CM})$$
$$= (51)\left(\frac{10}{17}\right)\left(1 - (0.0075)\,|36 - 30|\right)$$
$$\times \left(0.82 + \frac{1.8}{44}\right)\left(1 - (0.0032)(71)\right)(0.81)(1.00)$$
$$= 15 \ \text{lbf}$$

**The answer is (B).**

**56.** Per *NCEES Handbook* Reference Dose, the reference dose (RfD) defines the dose of a noncarcinogen that will produce an unacceptable risk to a population upon exposure.

**The answer is (D).**

**57.** The quality factor, $Q_F$, for alpha particles is 20. The dose equivalent is

$$h = \text{absorbed} = (0.13 \ \text{rad})(20) = 2.6 \ \text{rem}$$

**The answer is (D).**

**58.** The vapor concentration of the chemical is determined from the equation in *NCEES Handbook* Vaporization Rate ($Q_m$, mass/time) from a Liquid Surface.

$$Q_m = [MKA_S P^{\text{sat}}/(R_g T_L)]$$
$$\frac{Q_m}{KA_S} = \frac{MP^{\text{sat}}}{R_g T_L} = C_v$$

With mole fraction,

$$C_v = \frac{xMP^{\text{sat}}}{R_g T_L}$$

$$M = 106 \ \text{g/mol}$$
$$R_g = 82.1 \ \text{cm}^3 \cdot \text{atm/mol} \cdot \text{K}$$
$$T_L = 281 \text{K}$$
$$x = 0.018$$
$$P^{\text{sat}} = 0.0092 \ \text{atm}$$

$$C_v = \frac{xMP^{\text{sat}}}{R_g T_L}$$

$$= \frac{(0.018)(0.0092 \ \text{atm})\left(106 \ \frac{\text{g}}{\text{mol}}\right)\left(10^{-3} \ \frac{\text{kg}}{\text{g}}\right)}{\left(82.1 \ \frac{\text{cm}^3 \cdot \text{atm}}{\text{mol} \cdot \text{K}}\right)(281\text{K})\left(\frac{1 \ \text{m}^3}{10^6 \ \text{cm}^3}\right)}$$

$$= 0.00076 \ \text{kg/m}^3$$

The emission rate is the vent flow rate multiplied by $C_v$.

$$\left(20 \ \frac{\text{ft}^3}{\text{min}}\right)\left(0.00076 \ \frac{\text{kg}}{\text{m}^3}\right)\left(1440 \ \frac{\text{min}}{\text{d}}\right)\left(\frac{1 \ \text{m}^3}{35.3 \ \text{ft}^3}\right)$$
$$= 0.62 \ \text{kg/d}$$

**The answer is (D).**

**59.** The sum of the contaminant concentrations is

$$C_o = 1385 \ \text{ppb} = 1.385 \ \text{mg/L}$$

At equilibrium in lead-follow mode operation, $C_e$ is the influent concentration.

$$\frac{X}{M} = \frac{\text{mg chemical}}{\text{g GAC}}$$
$$= (2.837)\left(1.385 \ \frac{\text{mg}}{\text{L}}\right)^{0.431}$$
$$= \frac{3.26 \ \text{mg chemical}}{\text{g GAC}} \quad (3.3 \ \text{mg/g})$$

**The answer is (D).**

**60.** From *NCEES Handbook* Bioconcentration Factor BCF, the arsenic concentration in fish is

$$BCF = C_{org}/C$$
$$C_{org} = (BCF)C$$
$$= \left(44\ \frac{L}{kg}\right)\left(0.270\ \frac{mg}{L}\right)$$
$$= 12\ mg\ arsenic/kg\ fish$$

From the table "Exposure" in *NCEES Handbook* Reference Dose, for ingestion of contaminated foods,

$$CDI = \frac{(CF)(IR)(FI)(EF)(ED)}{(BW)(AT)}$$

For this situation, the equation reduces to

$$CDI = \frac{\left(\begin{array}{c}\text{concentration}\\ \text{in fish}\end{array}\right)(\text{daily IR})(ED)}{(BW)(\text{lifespan})}$$

$$= \frac{\left(12\ \frac{mg}{kg}\right)\left(54\ \frac{g}{d}\right)(1)\times(12\ yr)\left(\frac{1\ kg}{1000\ g}\right)}{(70\ kg)(70\ yr)}$$

$$= 1.6\times10^{-3}\ mg/kg\cdot d$$

From *NCEES Handbook* Carcinogens, the risk is

$$risk = (CDI)(CSF)$$
$$= \left(1.6\times10^{-3}\ \frac{mg}{kg\cdot d}\right)\left(1.75\left(\frac{mg}{kg\cdot d}\right)^{-1}\right)$$
$$= 2.8\times10^{-3}$$

**The answer is (B).**

**61.** From the table "Typical Values of $Rv$" in the *NCEES Handbook*, the $Rv$ value for medium-to-fine sand and gasoline is 80.

From *NCEES Handbook* Vadose Zone Penetration, the volume of gasoline spilled is

$$D = \frac{RvV}{A}$$
$$V = \frac{DA}{Rv}$$
$$= \frac{(10.3\ ft)(1160\ ft^2)\left(7.48\ \frac{gal}{ft^3}\right)}{80}$$
$$= 1117\ gal\quad(1100\ gal)$$

**The answer is (C).**

**62.** From *NCEES Handbook* First-Order Irreversible Reaction Kinetics, the mass transfer coefficient is

$$-\ln(C_A/C_{Ao}) = -kt$$
$$K_L a = \frac{-\ln\dfrac{C_A}{C_{Ao}}}{t}$$

At 20 s,

$$K_L a = \frac{-\ln\left(\dfrac{1620\ \frac{mg}{L}}{2000\ \frac{mg}{L}}\right)}{20\ s} = 0.0105\ s^{-1}$$

At 60 s,

$$K_L a = \frac{-\ln\left(\dfrac{1023\ \frac{mg}{L}}{2000\ \frac{mg}{L}}\right)}{60\ s} = 0.0112\ s^{-1}$$

At 240 s,

$$K_L a = \frac{-\ln\left(\dfrac{134\ \frac{mg}{L}}{2000\ \frac{mg}{L}}\right)}{240\ s} = 0.0113\ s^{-1}$$

At 360 s,

$$K_L a = \frac{-\ln\left(\dfrac{45\ \frac{mg}{L}}{2000\ \frac{mg}{L}}\right)}{360\ s} = 0.0105\ s^{-1}$$

The overall $K_L a$ is

$$K_L a = \frac{\frac{0.0105}{s} + \frac{0.0112}{s} + \frac{0.0113}{s} + \frac{0.0105}{s}}{4}$$
$$= 0.0109 \text{ s}^{-1} \quad (0.011 \text{ s}^{-1})$$

**The answer is (A).**

**63.** Adapted from *NCEES Handbook* Heats of Reaction,

$$\Delta G°_{\text{reaction}} = \sum \Delta G°_{\text{products}} - \sum \Delta G°_{\text{reactants}}$$
$$= (1 \text{ mol})\left(-108.1 \frac{\text{kcal}}{\text{mol}}\right)$$
$$- (1 \text{ mol})\left(-5.83 \frac{\text{kcal}}{\text{mol}}\right)$$
$$- (2 \text{ mol})\left(-37.6 \frac{\text{kcal}}{\text{mol}}\right)$$
$$= -27.07 \text{ kcal/mol}$$

The negative sign indicates the reaction proceeds as written. Therefore, sodium hydroxide will precipitate the lead.

**The answer is (B).**

**64.** From the table "Intake Rates" in the *NCEES Handbook*, the recommended values for estimating intake are

amount of water ingested, child, IR = 1.5 L/d
exposure frequency, EF = 365 d/yr
body weight for a child aged 6 yr to 12 yr, BW = 33 kg

From the table "Exposure" in *NCEES Handbook* Reference Dose, the chronic daily intake through ingestion with drinking water is

$$\text{CDI} = \frac{(\text{CW})(\text{IR})(\text{EF})(\text{ED})}{(\text{BW})(\text{AT})}$$
$$= \frac{\left(0.080 \frac{\text{mg}}{\text{L}}\right)\left(1.5 \frac{\text{L}}{\text{d}}\right)\left(365 \frac{\text{d}}{\text{yr}}\right)(7 \text{ yr})}{(33 \text{ kg})\left(365 \frac{\text{d}}{\text{yr}}\right)(70 \text{ yr})}$$
$$= 3.636 \times 10^{-4} \text{ mg/kg·d} \quad (3.6 \times 10^{-4} \text{ mg/kg·d})$$

**The answer is (B).**

**65.** From *NCEES Handbook* Specific Discharge, the average velocity of the light nonaqueous phase fuel is

$$q = -K(dh/dx)$$
$$v = q/n$$

Take $v$ as positive, so $K$ is positive.

$$v = \frac{K\left(\frac{dh}{dx}\right)}{n}$$
$$= \frac{\left(10 \frac{\text{m}}{\text{d}}\right)\left(0.0017 \frac{\text{m}}{\text{m}}\right)}{0.24}$$
$$= 0.071 \text{ m/d}$$

**The answer is (C).**

**66.** From *NCEES Handbook* Soil-Water Partition Coefficient $K_d = K_\rho$, the soil-water partition coefficient is

$$K_d = K_{oc} f_{oc}$$
$$= \left(364 \frac{\text{mL}}{\text{g}}\right)(0.17)$$
$$= 61.9 \text{ mL/g} \quad (62 \text{ mL/g})$$

**The answer is (B).**

**67.** From *NCEES Handbook* Time-Weighted Average (TWA), the time-weighted average is

$$\text{TWA} = \frac{\sum_{t=1}^{n} c_i t_i}{\sum_{i=1}^{n} t_i}$$
$$= \frac{c_1 t_1 + c_2 t_2}{t_1 + t_2}$$

| | | |
|---|---|---|
| TWA | time-weighted average | ppm |
| $c$ | concentration 1, 2 of chemical | ppm |
| $t$ | time of exposure 1, 2 to chemical | h |

$$\text{TWA} = \frac{c_1 t_1 + c_2 t_2}{t_1 + t_2}$$
$$= \frac{(12 \text{ ppm})(3 \text{ h}) + (4 \text{ ppm})(5 \text{ h})}{3 \text{ h} + 5 \text{ h}}$$
$$= 7 \text{ ppm}$$

**The answer is (C).**

**68.** Radon gas is emitted from water during normal household uses, but only when the water supply is untreated groundwater. Treated water poses little risk of radon exposure.

**The answer is (A).**

**69.** The material name on the safety data sheet (SDS) must always match the name printed on the material container.

**The answer is (D).**

**70.** From *NCEES Handbook* Safety Data Sheet (SDS), the SDS does not include the MCL, RMCL, or MCLG for the material. These parameters are associated with acceptable concentrations in drinking water and have no direct relationship with workplace safety.

**The answer is (B).**

**71.** From *NCEES Handbook* Predicting Lower Flammable Limits of Mixtures of Flammable Gases (Le Chatelier's Rule), the equation for the lower flammability limit (LFL) for the mixture is

$$\text{LFL}_{\text{mix}} = \frac{100}{\sum_{i=1}^{n}(C_{fi}/\text{LFL}_i)}$$

The LFL values for the gases in the mixture are found in *NCEES Handbook* Flammability.

| gas | gas volume in air (%) | LFL (%) |
|---|---|---|
| ethyl ether | 5.4 | 1.9 |
| ethylene | 7.1 | 2.7 |
| methane | 2.8 | 5 |

The LFL for the mixture is

$$\text{LFL}_{\text{mix}} = \frac{100}{\frac{5.4}{1.9} + \frac{7.1}{2.7} + \frac{2.8}{5}} = 16.6\% \quad (17\%)$$

**The answer is (C).**

**72.** From the table "Selected Chemical Interaction Effects" in the *NCEES Handbook*, additive effects do not increase substance toxicity—each substance contributes as if it were present without the other. Antagonistic effects occur when one substance cancels some of the negative effects of the other. "Neutral" is not used as a descriptor for chemical interaction effects. Synergistic effects occur when combined substances produce greater toxicity effects than either substance does when alone. Combined exposure to cigarette smoke and asbestos fibers is characterized as synergistic.

**The answer is (D).**

**73.** From *NCEES Handbook* Noise Pollution,

$$\Delta \text{SPL} = 10\log(r_1/r_2)^2$$
$$= 10\log\left(\frac{824 \text{ ft}}{1485 \text{ ft}}\right)^2$$
$$= 5.1 \text{ dB}$$

**The answer is (D).**

**74.** See *NCEES Handbook* Security. According to the U.S. Department of Homeland Security (DHS), risk-based performance standards for chemical facilities specify the goals chemical facilities must meet to gain DHS approval. The performance standards listed pertain to: the restriction of a facility's perimeter; the security of site assets; the screening and controlling of area access; attack deterrence, detection, and delay; hazardous material shipping, receipt, and storage; theft and the diversion of chemicals; sabotage; cyber-sabotage; emergency response plans; monitoring; training; personnel surety; elevated threats and protective measures; specific threats, vulnerabilities, or risks; the reporting of significant security incidents; significant security incidents and suspicious activities; officials and organization; and record keeping.

Resistance and defense are not included in the list of performance standards.

**The answer is (B).**

**75.** From *NCEES Handbook* Weir Formulas: Rectangular, the equation for the flow rate of suppressed free discharge is

$$Q = CLH^{3/2}$$

From *NCEES Handbook* Weir Formulas: V-Notch, for a rectangular weir, using SI units, the value of $C$ is 1.84. The flow rate is

$$Q = CLH^{3/2}$$
$$= (1.84)(1.6 \text{ m})(0.23 \text{ m})^{3/2}$$
$$= 0.32 \text{ m}^3/\text{s}$$

**The answer is (B).**

**76.** Assume all the $CO_2$ is used when carbonating. From *NCEES Handbook* Overall Coefficients,

$$C_A^* = p_{AG}/H$$

| $C_A^*$ | $CO_2$ concentration | (unknown) |
| $p_{AG}$ | partial pressure | 1.2 atm |

$$\begin{aligned} C_A^* &= p_{AG}/H \\ &= \dfrac{1.2 \text{ atm}}{1040 \dfrac{\text{atm}}{\text{mol fraction}}} \\ &= 0.00115 \text{ mol fraction} \end{aligned}$$

$$0.00115 \text{ mol fraction} = \dfrac{\text{mol } CO_2}{\text{mol } CO_2 + \text{mol } H_2O}$$

$$\text{mol } H_2O = \dfrac{1000 \dfrac{\text{g}}{\text{L}}}{16 \dfrac{\text{g}}{\text{mol}} + (2)\left(1 \dfrac{\text{g}}{\text{mol}}\right)}$$

$$= 55.6 \text{ mol/L}$$

In 1 L of carbonated water,

$$\dfrac{\text{mol } CO_2}{\text{L}} = 0.00115 \dfrac{\text{mol}}{\text{L}} + (0.00115)\left(55.6 \dfrac{\text{mol}}{\text{L}}\right)$$

$$= 0.065 \text{ mol/L}$$

$$\left(0.065 \dfrac{\text{mol}}{\text{L}}\right)\left(12 \dfrac{\text{g}}{\text{mol}} + (2)\left(16 \dfrac{\text{g}}{\text{mol}}\right)\right) = 2.9 \text{ g/L}$$

$$\dfrac{(100 \text{ kg})\left(1000 \dfrac{\text{g}}{\text{kg}}\right)}{2.9 \dfrac{\text{g}}{\text{L}}} = 34\,483 \text{ L/100 kg}$$

$$(35\,000 \text{ L/100 kg})$$

**The answer is (D).**

**77.** From *NCEES Handbook* Storm Return Period, the probability is

$$P = \dfrac{1}{T_r} = \dfrac{1}{15}$$

$$= 0.006667$$

From *NCEES Handbook* Normal Distribution, the distribution function is

$$W(x) = 0.5 - 0.006667$$

$$= 0.4333$$

From normal distribution tables, between 0 and $x$, $x = 1.50$.

The expected rainfall depth at 15 years is

$$48 \text{ in} + (1.50)(11.2 \text{ in}) = 64.8 \text{ in} \quad (65 \text{ in})$$

**The answer is (C).**

**78.** See *NCEES Handbook* CPM Precedence Relationships. When applying the critical path method (CPM) for project scheduling, float is the amount of extra time available either at the beginning or end of a task. This is determined by either late start–early start (LS–ES) or late finish–early finish (LF–EF). The time between a start, whether early or late, and a finish, whether early or late, is the task duration.

**The answer is (D).**

**79.** See *NCEES Handbook* Economics. $A$ is the recurring annual cost. The trend in the table shows cost increasing by $1000 each year, or $13,000 by the second year. Therefore, $A$ is $12,000, and $G$, the uniform increase in the annual cost each year, is $1000.

The first-year cost in excess of $G$ is $18,000 − $1000 = $17,000.

$$\begin{aligned} P &= A(P/A, 1.5\%, 10) + G(P/G, 1.5\%, 10) \\ &\quad + \$17{,}000 + P - F(P/F, 1.5\%, 10) \\ &= \$12{,}000(P/A, 1.5\%, 10) + \$1000(P/G, 1.5\%, 10) \\ &\quad + \$17{,}000 + \$2{,}300{,}000 - \$143{,}000(P/F, 1.5\%, 10) \end{aligned}$$

From interest tables,

$$(P/A, 1.5\%, 10) = 9.2222$$
$$(P/G, 1.5\%, 10) = 40.3675$$
$$(P/F, 1.5\%, 10) = 0.8617$$

Calculate the amount needed to finance the project.

$$\begin{aligned} P &= \$12{,}000(P/A, 1.5\%, 10) + \$1000(P/G, 1.5\%, 10) \\ &\quad + \$17{,}000 + \$2{,}300{,}000 - \$143{,}000(P/F, 1.5\%, 10) \\ &= (\$12{,}000)(9.2222) + (\$1000)(40.3675) \\ &\quad + \$17{,}000 + \$2{,}300{,}000 - (\$143{,}000)(0.8617) \\ &= \$2{,}708{,}120 \quad (\$2{,}700{,}000) \end{aligned}$$

**The answer is (D).**

**80.** The positions calculated by the GPS satellites are in the WGS84 UTM projection, and use the ellipsoidal elevation as defined by the WGS84 spheroid. The WGS84 datum is the accepted standard spheroid calculated using GPS satellites. The WGS84 Geographic datum is a geographic coordinate system that uses the WGS84 spheroid, but uses latitude and longitude, rather than northings and eastings. This is not used as the default datum for coordinates from GPS satellites. While the NAD27 and NAD83 datums are UTM coordinate systems using northings and eastings rather than latitudes and longitudes, these are older datums than WGS84 and use slightly different spheroids with different semi-major and semi-minor axis lengths.

*The answer is (A).*

# Solutions
# Morning Session 2

**81.**

| pipe no. | pipe diameter, $D$ (mm) | pipe length $L$ (km) | average flow, $Q_{ave}$ (L/d) | average flow, $Q'_{ave}$ (L/km·mm·d) | percent of I/I based on L/km·mm·d |
|---|---|---|---|---|---|
| 1 | 600 | 0.085 | 2500 | 49.0 | 26.7 |
| 2 | 600 | 0.046 | 2000 | 72.5 | 39.5 |
| 3 | 300 | 0.040 | 140 | 11.7 | 6.4 |
| 4 | 250 | 0.067 | 500 | 29.8 | 16.2 |
| 5 | 200 | 0.091 | 50 | 2.75 | 1.5 |
| 6 | 200 | 0.131 | 80 | 3.10 | 1.7 |
| 7 | 150 | 0.050 | 110 | 14.7 | 8.0 |
|   |   |   | total average flow | 183.55 |   |

The values in the table were calculated from

$$Q'_{ave} = \frac{Q_{ave}}{DL}$$

$$\text{percent of I/I based on L/km·mm·d} = \frac{Q'_{ave} \times 100\%}{183.55 \text{ L/km·mm·d}}$$

Pipes 1, 2, and 4 account for 82.4% of the measured I/I.

**The answer is (C).**

**82.** The BOD mass loading rate will remain unchanged since the population and wastewater generation activities remain unchanged. If only the flow decreases, then the concentration will increase and BOD mass loading will remain constant.

**The answer is (A).**

**83.** From the illustration, for a pre-to-post-development runoff ratio, $Q_o/Q_i$, of 0.42, the required storage-to-runoff volume ratio is

$$\frac{V_S}{V_R} = 0.32$$

**The answer is (B).**

**84.** From the graph, at a settling zone depth of $Z_o = 2.5$ m and a settling time of $t = 120$ min, $h_o = 77\%$.

Because $h_o = 77\% > h_o = 64\%$ when $t = 100$ min, the overall efficiency will be greater than 73%.

The efficiency increases.

**The answer is (B).**

**85.** Assume at 18% solids, the settled solids density is the same as water, 1000 kg/m³.

The volume of TSS removed at 18% solids is

$$\frac{(0.80)\left(12\,000\ \frac{m^3}{d}\right)\left(195\ \frac{mg}{L}\right)\left(10^{-6}\ \frac{kg}{mg}\right)}{\left(1000\ \frac{kg}{m^3}\right)(0.18)\left(\frac{1\ m^3}{10^3\ L}\right)}$$

$$= 10.4\ m^3/d \quad (10\ m^3/d)$$

**The answer is (C).**

**86.** From the table "Periodic Table of Elements" in the *NCEES Handbook*,

$$CrO_3\ MW = 52\ \frac{g}{mol} + (3)\left(16\ \frac{g}{mol}\right)$$
$$= 100\ g/mol \quad (100\ mg/mmol)$$

$$\frac{534\ \frac{mg}{L}}{100\ \frac{mg}{mmol}} = 5.34\ mmol/L$$

$$Cr_2(SO_4)_3\ MW = (2)\left(52\ \frac{g}{mol}\right) + (3)\left(\begin{array}{l}32\ \frac{g}{mol} + (4)\\ \times\left(16\ \frac{g}{mol}\right)\end{array}\right)$$
$$= 392\ g/mol \quad (392\ mg/mmol)$$

From the reduction reaction, $(4/4)(5.34\ mmol/L)$ of $CrO_3$ will react to produce $(2/4)(5.34\ mmol/L)$ of $Cr_2(SO_4)_3$.

$$\left(\frac{2}{4}\right)\left(5.34\ \frac{mmol}{L}\right) = 2.67\ mmol/L$$

The facility operates 24 h/d for 365 d/yr.

$$\left(392 \frac{\text{mg}}{\text{mmol}}\right)\left(2.67 \frac{\text{mmol}}{\text{L}}\right)\left(1.8 \frac{\text{m}^3}{\text{min}}\right)$$
$$\times \left(10^{-3} \frac{\text{kg} \cdot \text{L}}{\text{mg} \cdot \text{m}^3}\right)\left(1440 \frac{\text{min}}{\text{d}}\right)\left(365 \frac{\text{d}}{\text{yr}}\right)$$
$$= 9.9 \times 10^5 \text{ kg/yr} \quad [\text{dry mass}]$$

**The answer is (C).**

**87.** Assume the wasted wet sludge density is approximately that of water at 1000 kg/m³.

The volume of wasted sludge at 6% solids is

$$\frac{500 \frac{\text{kg}}{\text{d}}}{\left(1000 \frac{\text{kg}}{\text{m}^3}\right)(0.06)} = 8.3 \text{ m}^3/\text{d}$$

**The answer is (B).**

**88.** The mean cell residence time for design is

$$\theta_c = \theta_{c,\min}(\text{SF}) = (3.0 \text{ d})(2.5)$$
$$= 7.5 \text{ d}$$

From *NCEES Handbook* Activated Sludge,

$$X_A = \frac{\theta_c Y (S_0 - S_e)}{\theta (1 + k_d \theta_c)}$$

$$\frac{1}{\theta_c} = \frac{Y(S_0 - S)}{\theta X_A} - k_d$$

$$\frac{1}{7.5 \text{ d}} = \left(\frac{(0.5)\left(192 \frac{\text{mg}}{\text{L}} - 20 \frac{\text{mg}}{\text{L}}\right)}{\theta \left(2300 \frac{\text{mg}}{\text{L}}\right)} - \frac{0.05}{\text{d}}\right)\left(24 \frac{\text{h}}{\text{d}}\right)$$

$$\theta = 4.896 \text{ h} \quad (4.9 \text{ h})$$

**The answer is (A).**

**89.** From *NCEES Handbook* Clarifier, when settling velocity controls design,

$$v_o = Q/A_{\text{surface}}$$

The required overflow rate is

$$v_o = Q/A_{\text{surface}} = 1.34 \frac{\text{m}}{\text{h}}$$
$$= 1.34 \text{ m}^3/\text{m}^2 \cdot \text{h} \quad (1.3 \text{ m}^3/\text{m}^2 \cdot \text{h})$$

**The answer is (A).**

**90.** From *NCEES Handbook* Blowers, the equation for the power requirement for an aeration blower is

$$P_w = \frac{WRT_1}{Cne}\left[\left(\frac{P_2}{P_1}\right)^{0.283} - 1\right]$$

From the table "Gases Found in Wastewater at Standard Conditions (0°C, 1 atm)" in the *NCEES Handbook*, the specific weight of air = 0.0808 lbf/ft³. The weight of flow of air is

$$W = \left(100 \frac{\text{ft}^3}{\text{sec}}\right)\left(0.0808 \frac{\text{lbf}}{\text{ft}^3}\right) = 8.08 \text{ lbf/sec}$$

From *NCEES Handbook* Temperature Conversions, °R = 73°F + 459.69° = 532.69°R

From the table "Conversion Factors" in the *NCEES Handbook*, 1.0 atm = 14.70 lbf/in².

The maximum allowable ambient pressure is

$$(1.3 \text{ atm})\left(14.70 \frac{\frac{\text{lbf}}{\text{in}^2}}{\text{atm}}\right) = 19.11 \text{ lbf/in}^2$$

The required power is

$$P_w = \frac{WRT_1}{Cne}\left[\left(\frac{P_2}{P_1}\right)^{0.283} - 1\right]$$

$$= \frac{\left(8.08 \frac{\text{lbf}}{\text{sec}}\right)\left(53.3 \frac{\text{ft-lbf}}{\text{lbf-°R}}\right)(532.69°\text{R})}{\left(550 \frac{\text{ft-lbf}}{\text{sec-hp}}\right)(0.283)(0.8)}$$

$$\times \left(\left(\frac{19.11 \frac{\text{lbf}}{\text{in}^2}}{14.70 \frac{\text{lbf}}{\text{in}^2}}\right)^{0.283} - 1\right)$$

$$= 141.9 \text{ hp} \quad (140 \text{ hp})$$

**The answer is (C).**

**91.**

| $t_i$ (min) | $C_i$ ($\mu$g/L) | $C_i \Delta t$ (min·$\mu$g/L) | $C_i \Delta t t_i$ (min²·$\mu$g/L) |
|---|---|---|---|
| 20 | 0 | 0 | 0 |
| 30 | 100 | 1000 | 30 000 |
| 40 | 390 | 3900 | 156 000 |
| 50 | 148 | 1480 | 74 000 |
| 60 | 83 | 830 | 49 800 |
| 70 | 47 | 470 | 32 900 |
| 80 | 25 | 250 | 20 000 |
| 90 | 12 | 120 | 10 800 |
| 100 | 7.5 | 75 | 7500 |
| 110 | 2.5 | 25 | 2750 |
|  |  | 8150 | 383 750 |

$C_i$ tracer concentration at $t_i$   $\mu$g/L

$\Delta t$ time interval between time measurements   min

$$\Delta t = 10 \text{ min}$$

$$t_a = \frac{\sum C_i \Delta t t_i}{\sum C_i \Delta t} = \frac{383\,750 \frac{\text{min}^2 \cdot \mu\text{g}}{\text{L}}}{8150 \frac{\text{min} \cdot \mu\text{g}}{\text{L}}}$$

$$= 47 \text{ min}$$

**The answer is (B).**

**92.** From *NCEES Handbook* Rapid Mix and Flocculator Design,

$$G = \sqrt{\frac{P}{\mu V}}$$

$$P = G^2 \mu V$$

From the table "Properties of Water (SI Metric Units)" in the *NCEES Handbook*, the water viscosity is $\mu = 0.001002$ Pa·s at 20°C assumed temperature. From *NCEES Handbook* Clarifier,

$$V = Qt = \left(5000 \frac{\text{m}^3}{\text{d}}\right)(120 \text{ s})\left(\frac{1 \text{ d}}{86\,400 \text{ s}}\right)$$

$$= 6.94 \text{ m}^3$$

$$P = \left(\frac{850}{\text{s}}\right)^2 (6.94 \text{ m}^3)(0.001002 \text{ Pa·s})\left(\frac{\text{N}}{\text{m}^2 \cdot \text{Pa}}\right)$$

$$\times \left(\frac{1 \text{ kW·s}}{1000 \text{ N·m}}\right)$$

$$= 5.0 \text{ kW}$$

**The answer is (B).**

**93.** From *NCEES Handbook* NRCS (SCS) Rainfall-Runoff, the maximum basin retention is

$$S = \frac{1,000}{CN} - 10$$

$$= \frac{1000}{49} - 10$$

$$= 10.4 \text{ in}$$

The runoff is

$$Q = \frac{(P - 0.2S)^2}{P + 0.8S}$$

$$= \frac{(3.6 \text{ in} - (0.2)(10.4 \text{ in}))^2}{3.6 \text{ in} + (0.8)(10.4 \text{ in})}$$

$$= 0.19 \text{ in}$$

**The answer is (B).**

**94.** From *NCEES Handbook* Rational Formula, the pre-development discharge of a 2 yr storm is

$$Q = CIA$$

$$= (0.25)\left(5.3 \frac{\text{cm}}{\text{h}}\right)\left(\frac{1 \text{ m}}{100 \text{ cm}}\right)\left(\frac{1 \text{ h}}{3600 \text{ s}}\right)$$

$$\times (100 \text{ ha})\left(10\,000 \frac{\text{m}^2}{\text{ha}}\right)$$

$$= 3.7 \text{ m}^3/\text{s}$$

**The answer is (C).**

**95.** From *NCEES Handbook* Needed Fire Flow Formula, for a Class 3 structure, $F = 0.8$.

$$C_i = 18F(A_i)^{0.5}$$

$$= (18)(0.8)\sqrt{9200 \text{ ft}^2}$$

$$= 1381$$

The needed fire flow is

$$\text{NFF} = (C_i)(O_i)[1.0 + (X+P)_i]$$

$$= (1381)(0.8)(1.0 + 0.23)$$

$$= 1359 \text{ gpm} \quad (1400 \text{ gpm})$$

**The answer is (B).**

**96.**

$$\left(400 \frac{\text{m}^3}{\text{d}}\right)\left(2000 \frac{\text{mg}}{\text{L}}\right)\left(10^{-6} \frac{\text{kg}}{\text{mg}}\right)\left(10^3 \frac{\text{L}}{\text{m}^3}\right)$$

$$= 800 \text{ kg BOD/d}$$

The pond 1 volume is

$$\frac{800 \frac{\text{kg BOD}}{\text{d}}}{0.35 \frac{\text{kg BOD}}{\text{m}^3 \cdot \text{d}}} = 2286 \text{ m}^3$$

The minimum depth for pond 1 is

$$\frac{2286 \text{ m}^3}{1000 \text{ m}^2} = 2.286 \text{ m} \quad (2.3 \text{ m})$$

**The answer is (B).**

**97.** Options A, B, and D would result in improved clarifier performance. Option C, decreasing the depths of the inlet baffles, would likely contribute to hydraulic short-circuiting and result in less efficient clarifier performance.

**The answer is (C).**

**98.** From the table "Common Radicals in Water" in the *NCEES Handbook*,

| ion | mg/L | MW (mg/mmol) | mmol/L |
|---|---|---|---|
| $SO_4^{-2}$ | 28 | 96 | 0.29 |
| $Ca^{+2}$ | 67 | 40 | 1.675 |
| $HCO_3^-$ | 353 | 61 | 5.79 |

From *NCEES Handbook* Lime-Soda Ash Softening Equations,

$$Ca^{+2} + 2HCO_3^- + Ca(OH)_2 \rightarrow 2CaCO_{3(S)}$$

The chemical produced is

$$(2)\left(1.675 \frac{\text{mmol}}{\text{L}} \text{ CaCO}_{3(S)}\right) = 3.35 \frac{\text{mmol}}{\text{L}} \text{ CaCO}_{3(S)}$$

$$HCO_3^- \text{ remaining} = 5.79 \frac{\text{mmol}}{\text{L}} - (2)\left(1.675 \frac{\text{mmol}}{\text{L}}\right)$$
$$= 2.44 \text{ mmol/L} > 0 \quad [\text{OK}]$$

The $CaCO_3$ sludge production is

$$\left(3.35 \frac{\text{mmol}}{\text{L}}\right)\left(100 \frac{\text{mg}}{\text{mmol}}\right) = 335 \text{ mg/L}$$

$$\left(335 \frac{\text{mg}}{\text{L}}\right)\left(500 \frac{\text{gal}}{\text{min}}\right)\left(3.785 \frac{\text{L}}{\text{gal}}\right)$$
$$\times \left(10^{-6} \frac{\text{kg}}{\text{mg}}\right)\left(1440 \frac{\text{min}}{\text{d}}\right) = 912 \text{ kg/d}$$

Assume sludge density is the same as water, 1000 kg/m³.

The sludge volume is

$$\frac{912 \frac{\text{kg}}{\text{d}}}{\left(1000 \frac{\text{kg}}{\text{m}^3}\right)(0.35)} = 2.6 \text{ m}^3/\text{d}$$

**The answer is (C).**

**99.** The dry mass TSS removed is

$$(0.80)\left(10\,000 \frac{\text{m}^3}{\text{d}}\right)\left(234 \frac{\text{mg}}{\text{L}}\right)\left(10^{-6} \frac{\text{kg}}{\text{mg}}\right)$$
$$\times \left(10^3 \frac{\text{L}}{\text{m}^3}\right) = 1872 \text{ kg/d} \quad [\text{dry}]$$

Assuming a solids density equal to that of water of 1000 kg/m³, the volume at 30% solids is

$$\frac{1872 \frac{\text{kg}}{\text{d}}}{\left(1000 \frac{\text{kg}}{\text{m}^3}\right)(0.30)} = 6.24 \text{ m}^3/\text{d} \quad (6.2 \text{ m}^3/\text{d})$$

**The answer is (C).**

**100.** From the illustration, the daily demand is

$$\frac{(14.9 \text{ MG} - 0 \text{ MG})\left(24 \frac{\text{hr}}{\text{day}}\right)}{2400 \text{ hr} - 0000 \text{ hr}} = 14.9 \text{ MGD} \quad (15 \text{ MGD})$$

**The answer is (D).**

**101.** The standard titrant used for alkalinity is 0.02N $H_2SO_4$ because 1 mL of this acid neutralizes 1 mg of alkalinity at $CaCO_3$. Therefore, because 0.03N $H_2SO_4$ is 1.5 times as concentrated as 0.02N $H_2SO_4$, 1 mL of 0.03N acid will neutralize 1.5 mg alkalinity as $CaCO_3$. This relationship is derived as follows.

From the table "Common Radicals in Water" in the *NCEES Handbook*, the equivalent weight of $CaCO_3$ is 50 g/equiv.

$$0.03N\ H_2SO_4 = 0.03\ \frac{equiv}{L}$$
$$= 0.03\ meq/mL$$
$$\left(0.03\ \frac{meq}{mL}\right)\left(50\ \frac{mg}{meq}\right) = 1.5\ mg\ as\ CaCO_3/mL$$

The total alkalinity is

$$\frac{(titrant\ volume)(1.5\ mg\ alkalinity\ as\ CaCO_3)}{(sample\ volume)(1.0\ mL\ titrant)}$$

$$= \frac{(28\ mL)(1.5\ mg\ alkalinity\ as\ CaCO_3) \times \left(1000\ \frac{mL}{L}\right)}{(200\ mL)(1.0\ mL)}$$

$$= 210\ mg/L\ as\ CaCO_3$$

**The answer is (C).**

**102.**

$$noncarbonate\ hardness$$
$$= total\ hardness - carbonate\ hardness$$

In this case, carbonate hardness is equal to total hardness; noncarbonate hardness is equal to 0 mg/L.

**The answer is (A).**

**103.** From the illustration, the minimum dose required to obtain a free chlorine residual is about 7.7 mg/L.

**The answer is (C).**

**104.** The annual total Kjeldahl nitrogen loading to the pond is

$$\frac{\left(2.23\ \frac{ft^3}{min}\right)\left(1.43\ \frac{mg}{L}\right)\left(28.25\ \frac{L}{ft^3}\right) \times \left(\frac{2.204\ lbm}{10^6\ mg}\right)\left(1440\ \frac{min}{day}\right)}{(24\ ac)(8\ ft)\left(\frac{1\ yr}{365\ days}\right)}$$
$$= 0.54\ lbm/ac\text{-}ft\text{-}yr$$

**The answer is (B).**

**105.** The four primary factors used to control incineration efficiency are burn temperature, oxygen level, turbulence, and residence time.

**The answer is (A).**

**106.** Interpolating from the table "Properties of Water (English Units)" in the *NCEES Handbook*, at 36°F, $P_g = 0.105$ psi.

From *NCEES Handbook* Psychrometrics, the relative humidity is

$$\Phi = P_v/P_g$$
$$= \frac{0.089\ \frac{lbf}{in^2}}{0.105\ \frac{lbf}{in^2}}$$
$$= 0.848\quad(85\%)$$

**The answer is (C).**

**107.** From *NCEES Handbook* Electrostatic Precipitator, the plate area is determined using

$$\eta = 1 - \exp(-Aw/Q)$$
$$0.8 = 1 - \exp\left(\frac{-A\left(1.0\ \frac{m}{s}\right)}{14\ \frac{m^3}{s}}\right)$$
$$A = 22.253\ m^2\quad(23\ m^2)$$

**The answer is (C).**

**108.** Applying Stokes' law from *NCEES Handbook* Stokes' Law, the particle settling velocity is

$$v_t = \frac{g(\rho_p - \rho_f)d^2}{18\mu}$$

$$= \frac{\left(9.81\ \frac{m}{s^2}\right)\left(0.58\ \frac{g}{cm^3} - 0.00152\ \frac{g}{cm^3}\right) \times (2.5\ \mu m)^2\left(100\ \frac{cm}{m}\right)^4\left(10^{-6}\ \frac{m}{\mu m}\right)^2}{(18)\left(1.9 \times 10^{-5}\ \frac{kg}{m \cdot s}\right)\left(10^3\ \frac{g}{kg}\right)}$$

$$= 0.0104\ cm/s\quad(0.010\ cm/s)$$

**The answer is (B).**

**109.** From *NCEES Handbook* Incineration, using the equation for the destruction and removal efficiency (DRE), the maximum allowable mass emission rate is

$$DRE = \frac{W_{in} - W_{out}}{W_{in}} \times 100\%$$

$$W_{out} = W_{in}\left(1 - \frac{DRE}{100\%}\right)$$

$$= \left(1000 \ \frac{kg}{h}\right)\left(1 - \frac{99.99\%}{100\%}\right)$$

$$= 0.10 \ kg/hr$$

**The answer is (B).**

**110.** Wind originating from the west-southwest exceeds 10 mph approximately 9% of the time.

**The answer is (C).**

**111.** The Environmental Protection Agency (EPA) has traditionally employed a command-and-control approach to implementing the Clean Air Act.

**The answer is (A).**

**112.** Photochemical smog is produced from a variety of complex physical and chemical reactions involving nitrogen oxides, carbon monoxide, hydrocarbons, sunlight, and many other factors combined under favorable conditions. Among the photochemical oxidants involved in photochemical smog formation, ozone is the most common and is an irritant to the mucosa and lungs. Formaldehyde and acrolein are common aldehyde constituents of photochemical smog that are associated with eye irritation. Therefore, the primary nuisance components of photochemical smog include ozone and oxidized hydrocarbons, particularly aldehydes.

**The answer is (C).**

**113.** From *NCEES Handbook* Oxidation Chemistry, the reaction equation for stoichiometric oxidation of a hydrocarbon in air is

$$C_xH_y + (b)O_2 + 3.76(b)N_2 \rightarrow xCO_2 + \left(\frac{y}{2}\right)H_2O + 3.76(b)N_2$$

For octane ($C_8H_{18}$), $x = 8$. Therefore, 8 moles of carbon dioxide are produced for every mole of $C_8H_{18}$ combusted.

Using the values given in the table "Periodic Table of Elements" in the *NCEES Handbook*, the molecular weight of carbon dioxide is

$$MW_{CO_2} = 12 \ \frac{g}{mol} + (2)\left(16 \ \frac{g}{mol}\right) = 44 \ g/mol$$

The molecular weight of $C_8H_{18}$ is

$$MW_{C_8H_{18}} = (8)\left(12 \ \frac{g}{mol}\right) + (18)\left(1 \ \frac{g}{mol}\right) = 114 \ g/mol$$

The mass of the $C_8H_{18}$ combusted is

$$(100 \ gal)\left(3.785 \ \frac{L}{gal}\right)\left(701 \ \frac{g}{L}\right) = 265\,329 \ g$$

The number of moles of $C_8H_{18}$ combusted is

$$\frac{265\,329 \ g}{114 \ \frac{g}{mol}} = 2327 \ mol$$

The number of moles of carbon dioxide combusted is

$$(2327 \ mol \ C_8H_{18})\left(8 \ \frac{mol \ CO_2}{mol \ C_8H_{18}}\right) = 18\,616 \ mol \ CO_2$$

The mass of carbon dioxide combusted is

$$\frac{(18\,616 \ mol \ CO_2)\left(44 \ \frac{g}{mol}\right)}{1000 \ \frac{g}{kg}} = 819 \ kg \quad (820 \ kg)$$

**The answer is (C).**

**114.** Increasing the cross-sectional area will decrease the velocity of the gas, $v_g$, causing a decrease in efficiency.

**The answer is (C).**

**115.** The location of the maximum ground-level concentration of pollutants emitted from the stack will occur along the plume centerline.

From the table "Atmospheric Stability" in the *NCEES Handbook*, for a wind speed of 5.3 m/s measured 10 m above ground level and with low incoming solar radiation, the stability category is D.

From *NCEES Handbook* Atmospheric Dispersion Modeling, for emissions from a stack with an effective height $H$, the standard deviation along the vertical axis under neutral atmospheric conditions is

$$\sigma_z = \frac{H}{\sqrt{2}} = (0.707)(120 \ m)$$

$$= 84.84 \quad (85)$$

From the figure "Vertical Standard Deviations of a Plume" in the *NCEES Handbook*, for atmospheric stability category B and a standard deviation along the vertical axis of 85, the maximum ground-level concentration of pollutants emitted from the stack occurs at 400 m.

**The answer is (B).**

**116.** A non-attainment area is defined as a region where the National Ambient Air Quality Standards (NAAQS) cannot be met.

**The answer is (B).**

**117.** The area per bag is

$$\begin{aligned}\pi(\text{bag diameter})\\ \times(\text{bag length})\end{aligned} = \pi(15 \text{ cm})\left(\frac{1 \text{ m}}{100 \text{ cm}}\right)(2.5 \text{ m})$$
$$= 1.18 \text{ m}^2/\text{bag}$$

The number of bags per 100 m² of total bag area is

$$\frac{100 \text{ m}^2}{1.18 \frac{\text{m}^2}{\text{bag}}} = 84.75 \text{ bags} \quad (85 \text{ bags})$$

**The answer is (A).**

**118.** From *NCEES Handbook* Cyclone Effective Number of Turns Approximation,

$$N_e = \frac{1}{H}\left[L_b + \frac{L_c}{2}\right]$$
$$= \left(\frac{1}{(0.15)(4.5 \text{ ft})}\right)\left((1.5)(4.5 \text{ ft}) + \frac{(2.5)(4.5 \text{ ft})}{2}\right)$$
$$= 18.3 \quad (18)$$

**The answer is (B).**

**119.** Conditions depicted by Illustration III are neutral, those shown by illustrations II and IV are stable, and those shown by Illustration I are unstable.

**The answer is (C).**

**120.** From the figure "Vertical Standard Deviations of a Plume" in the *NCEES Handbook*, the standard deviation along the vertical axis for a 3800 m distance is $\sigma_z = 200$ m.

From the figure "Horizontal Standard Deviations of a Plume" in the *NCEES Handbook*, the standard deviation along the horizontal axis for a 3800 m distance is $\sigma_y = 380$ m.

| | | |
|---|---|---|
| $Q$ | emission rate | 26 kg/s |
| $\mu$ | wind speed | 3.2 m/s |

From *NCEES Handbook* Atmospheric Dispersion Modeling (Gaussian), the maximum concentration at ground level and directly downwind from an elevated source is

$$C_{3800,0} = \frac{Q}{\pi u \sigma_z \sigma_y} \exp\left(-\tfrac{1}{2}(H^2/\sigma_z^2)\right)$$
$$= \frac{\left(26 \frac{\text{kg}}{\text{s}}\right)\left(10^6 \frac{\text{mg}}{\text{kg}}\right) e^{(-0.5)(110 \text{ m}/200 \text{ m})^2}}{\pi\left(3.2 \frac{\text{m}}{\text{s}}\right)(200 \text{ m})(380 \text{ m})}$$
$$= 29.3 \text{ mg/m}^3 \quad (29 \text{ mg/m}^3)$$

**The answer is (B).**

# Solutions
## Afternoon Session 2

**121.** If a 1-to-1 molar ratio is assumed, 1 mol of sodium hydroxide (NaOH) will neutralize 1 mol of hydrochloric acid (HCl). Therefore, 2.1 mol of NaOH is needed per liter of 2.1 molar HCl treated. From the table "Periodic Table of the Elements" in the *NCEES Handbook*,

$$\text{MW NaOH} = 23 \ \frac{\text{g}}{\text{mol}} + 16 \ \frac{\text{g}}{\text{mol}} + 1 \ \frac{\text{g}}{\text{mol}}$$
$$= 40 \ \text{g/mol}$$
$$= \left(2.1 \ \frac{\text{mol}}{\text{L}}\right)\left(40 \ \frac{\text{g}}{\text{mol}}\right)\left(300\,000 \ \frac{\text{L}}{\text{d}}\right)$$
$$\times \left(\frac{1 \ \text{kg}}{1000 \ \text{g}}\right)$$
$$= 25\,200 \ \text{kg/d} \quad (25\,000 \ \text{kg/d})$$

**The answer is (D).**

**122.** Assume a 100 kg sample.

| waste component | discarded mass (kg) | dry mass (kg) |
|---|---|---|
| food | 13 | 3.9 |
| glass | 6 | 5.9 |
| plastic | 4 | 3.9 |
| paper | 37 | 35 |
| cardboard | 10 | 9.5 |
| textiles | 1 | 0.9 |
| ferrous metal | 8 | 7.8 |
| nonferrous metal | 2 | 2.0 |
| wood | 4 | 3.2 |
| yard clippings | 15 | 6.0 |
|  | 100 | 78.1 |

$$\text{discarded mass, kg} = (100 \ \text{kg})\left(\frac{\% \ \text{mass}}{100}\right)$$

$$\text{dry mass, kg} = \frac{(\text{discarded mass, kg}) \times (100 - \% \ \text{moisture})}{100}$$

$$(100 \ \text{kg} - 78.1 \ \text{kg})\left(\frac{100\%}{100 \ \text{kg}}\right) = 21.9\% \quad (22\%)$$

**The answer is (B).**

**123.** Over 90% of the gas volume produced from anaerobic decomposition of solid waste in landfills consists of methane ($CH_4$) and carbon dioxide ($CO_2$).

**The answer is (B).**

**124.** The containment volume based on the largest tank is

$$(1{,}000{,}000 \ \text{gal})(1.1)\left(0.134 \ \frac{\text{ft}^3}{\text{gal}}\right) = 147{,}400 \ \text{ft}^3$$

The containment volume based on stormwater is

$$(1.85 \ \text{ac})(14 \ \text{in})\left(43{,}560 \ \frac{\text{ft}^2}{\text{ac}}\right)\left(\frac{1 \ \text{ft}}{12 \ \text{in}}\right) = 94{,}017 \ \text{ft}^3$$

The berm height without freeboard is

$$\frac{147{,}400 \ \text{ft}^3 + 94{,}017 \ \text{ft}^3}{(1.85 \ \text{ac})\left(43{,}560 \ \frac{\text{ft}^2}{\text{ac}}\right)} = 3.0 \ \text{ft}$$

The berm height with freeboard is

$$3.0 \ \text{ft} + 1.5 \ \text{ft} = 4.5 \ \text{ft}$$

**The answer is (C).**

**125.** Assume that the sludge density is equal to the water density (1000 kg/m³).

$$\text{sludge mass} = (\text{volume})(\text{density})$$
$$\times (1 - \text{fractional moisture})$$
$$= \left(4 \ \frac{\text{m}^3}{\text{d}}\right)\left(1000 \ \frac{\text{kg}}{\text{m}^3}\right)(1 - 0.76)\left(\frac{1 \ \text{d}}{24 \ \text{h}}\right)$$
$$= 40 \ \text{kg/h}$$

The hourly heat value of the dry sludge solids is

$$\left(40 \ \frac{\text{kg}}{\text{h}}\right)\left(30\,000 \ \frac{\text{kJ}}{\text{kg}}\right) = 1.2 \times 10^6 \ \text{kJ/h}$$

**The answer is (B).**

**126.** From *NCEES Handbook* Break-Through Time for Leachate to Penetrate a Landfill Clay Liner, the break-through time is

$$t = \frac{d^2\eta}{K(d+h)}$$

$$= \frac{(500 \text{ mm})^2(0.2)\left(10^{-3}\frac{\text{m}}{\text{mm}}\right)^2}{\left(10^{-9}\frac{\text{m}}{\text{s}}\right)\left(86\,400\frac{\text{s}}{\text{d}}\right)\left(365\frac{\text{d}}{\text{yr}}\right)}$$

$$\times \left(500 \text{ mm} + 140 \text{ mm}\right)\left(10^{-3}\frac{\text{m}}{\text{mm}}\right)$$

$$= 2.48 \text{ yr} \quad (2.6 \text{ yr})$$

**The answer is (B).**

**127.** The reaction between hydrogen cyanide and sodium hydroxide is

$$HCN + NaOH \rightarrow NaCN + H_2O$$

From the table "Periodic Table of Elements" in the *NCEES Handbook*,

$$\text{MW HCN} = 1\frac{\text{g}}{\text{mol}} + 12\frac{\text{g}}{\text{mol}} + 14\frac{\text{g}}{\text{mol}}$$

$$= 27 \text{ g/mol}$$

$$\left(1.94\frac{\text{g}}{\text{m}^3}\right)\left(1.9\frac{\text{m}^3}{\text{s}}\right) = 3.69 \text{ g/s}$$

$$\frac{3.69\frac{\text{g}}{\text{s}}}{27\frac{\text{g}}{\text{mol}}} = 0.137 \text{ mol/s}$$

0.137 mol/s of HCN will react with 0.137 mol/s NaOH. The sodium hydroxide flow rate is

$$\frac{\left(0.137\frac{\text{mol}}{\text{s}}\right)\left(60\frac{\text{s}}{\text{min}}\right)\left(1\frac{\text{equiv}}{\text{mol}}\right)}{0.1\frac{\text{equiv}}{\text{L}}} = 82 \text{ L/min}$$

**The answer is (D).**

**128.** Assume a 100 kg sample.

| component | mass (kg) | discarded energy (kJ/kg) | discarded energy (kJ) |
|---|---|---|---|
| food | 16 | 4650 | 74 400 |
| glass | 6 | 150 | 900 |
| plastic | 5 | 32 600 | 163 000 |
| paper | 44 | 16 750 | 737 000 |
| ferrous metal | 5 | 840 | 4200 |
| nonferrous metal | 5 | 700 | 3500 |
| yard clippings | 19 | 6500 | 123 500 |
| | 100 | | 1 106 500 |

discarded energy, kJ
$$= (\text{mass, kg})(\text{discarded energy, kJ/kg})$$

The discarded bulk waste energy content is

$$\frac{1\,106\,500 \text{ kJ}}{100 \text{ kg}} = 11\,065 \text{ kJ/kg} \quad (11\,000 \text{ kJ/kg})$$

**The answer is (C).**

**129.** The molar concentration of $Pb^{+2}$ is

$$\frac{\left(72\frac{\text{mg}}{\text{L}}\right)\left(10^{-3}\frac{\text{g}}{\text{mg}}\right)}{207\frac{\text{g}}{\text{mol}}} = 0.000348 \text{ mol/L}$$

The molar concentration of $OH^-$ at pH 7.6 is by definition

$$pH + pOH = 14$$
$$pOH = 14 - 7.6$$
$$= 6.4$$

From *NCEES Handbook* Acids, Bases, and pH,

$$pH = \log_{10}\left(\frac{1}{[H^+]}\right)$$

Therefore,

$$\text{pOH} = \log_{10}\left(\frac{1}{[\text{OH}^-]}\right)$$
$$[\text{OH}^-] = 10^{-\text{pOH}}$$
$$[\text{OH}^-] = 10^{-6.4}\,\frac{\text{mol}}{\text{L}}$$
$$= 3.98 \times 10^{-7}\,\text{mol/L}$$
$$[\text{PbOH}^+] = (10^{7.82})\left(0.000348\,\frac{\text{mol}}{\text{L}}\right)$$
$$\times \left(3.98 \times 10^{-7}\,\frac{\text{mol}}{\text{L}}\right)$$
$$= 0.00915\,\text{mol/L}$$

$$\left(0.00915\,\frac{\text{mol}}{\text{L}}\right)\left(207\,\frac{\text{g}}{\text{mol}}\right)\left(1000\,\frac{\text{mg}}{\text{g}}\right)$$
$$= 1894\,\text{mg/L as Pb}^{+2}$$

$$[\text{Pb(OH)}_2] = (k_2)[\text{pbOH}^+][\text{OH}^-]$$
$$= (10^{3.03})\left(0.00915\,\frac{\text{mol}}{\text{L}}\right)$$
$$\times \left(3.98 \times 10^{-7}\,\frac{\text{mol}}{\text{L}}\right)$$
$$= 3.9 \times 10^{-6}\,\text{mol/L}$$

$$\left(3.9 \times 10^{-6}\,\frac{\text{mol}}{\text{L}}\right)\left(207\,\frac{\text{g}}{\text{mol}}\right)\left(1000\,\frac{\text{mg}}{\text{g}}\right)$$
$$= 0.808\,\text{mg/L as Pb}^{+2}$$

$$[\text{Pb(OH)}_3^-] = (k_3)[\text{Pb(OH)}_2][\text{OH}^-]$$
$$= (10^{3.73})\left(3.9 \times 10^{-6}\,\frac{\text{mol}}{\text{L}}\right)$$
$$\times \left(3.98 \times 10^{-7}\,\frac{\text{mol}}{\text{L}}\right)$$
$$= 8.3 \times 10^{-9}\,\text{mol/L}$$

$$\left(8.3 \times 10^{-9}\,\frac{\text{mol}}{\text{L}}\right)\left(207\,\frac{\text{g}}{\text{mol}}\right)\left(1000\,\frac{\text{mg}}{\text{g}}\right)$$
$$= 0.0017\,\text{mg/L as Pb}^{+2}$$

At pH 7.6, the dominant $\text{Pb}^{+2}$ specie is $\text{PbOH}^+$.

**The answer is (B).**

**130.** The total as-discarded waste volume to be collected is

$$\frac{(1400\,\text{people})\left(3.6\,\frac{\text{lbm}}{\text{day·person}}\right)}{240\,\frac{\text{lbm}}{\text{yd}^3}} = 21\,\text{yd}^3/\text{day}$$

Assume waste collection occurs once each week. The compacted volume per week is

$$\frac{\left(21\,\frac{\text{yd}^3}{\text{day}}\right)\left(7\,\frac{\text{days}}{\text{wk}}\right)}{\left(\frac{3.1\,\text{yd}^3\,\text{discarded}}{1\,\text{yd}^3\,\text{compacted}}\right)} = 47.4\,\text{yd}^3/\text{wk}$$

The number of truck loads to collect all the waste in one week is

$$\frac{47.4\,\frac{\text{yd}^3}{\text{wk}}}{10\,\frac{\text{yd}^3}{\text{load}}} = 4.74\,\text{loads/wk} \quad (5\,\text{loads/wk})$$

The time required to fill one truck is

$$\frac{\left(\frac{32\,\text{sec} + 18\,\text{sec}}{\text{residence}}\right)\left(10\,\frac{\text{yd}^3}{\text{load}}\right)}{\left(\frac{47.4\,\text{yd}^3}{388\,\text{residences}}\right)\left(60\,\frac{\text{sec}}{\text{min}}\right)} = 68\,\text{min}$$

The available collection time is

$$\left(8\,\frac{\text{hr}}{\text{day}}\right)\left(60\,\frac{\text{min}}{\text{hr}}\right) = 480\,\text{min/day}$$

At the beginning of day 1,

| activity | incremental time (min) | cumulative time (min) |
|---|---|---|
| travel to route | 28 | 28 |
| collect load 1 | 68 | 96 |
| travel to landfill | 47 | 149 |
| unload | 15 | 158 |
| travel to route | 47 | 205 |
| collect load 2 | 68 | 273 |
| travel to landfill | 47 | 320 |
| unload | 15 | 335 |
| travel to route | 47 | 382 |
| collect load 3 | 68 | 450 |
| travel to yard | 28 | 478 |

Compared to the total time available of 480 min, the cumulative time for all activities of 478 min is 2 min early.

At the beginning of day 2,

| activity | incremental time (min) | cumulative time (min) |
|---|---|---|
| travel to landfill | 34 | 34 |
| unload | 15 | 49 |
| travel to route | 47 | 96 |
| collect load 4 | 68 | 164 |
| travel to landfill | 47 | 211 |
| unload | 15 | 226 |
| travel to route | 47 | 273 |
| collect load 5 | 68 | 341 |
| travel to landfill | 47 | 388 |
| unload | 15 | 403 |
| travel to yard | 34 | 437 |

Compared to the total time available of 480 min, the cumulative time for all activities of 437 min is 43 min early.

The number of days required to collect all the waste is two.

**The answer is (B).**

**131.** The annual waste volume to be landfilled is

$$\frac{(1-0.24)(117{,}000 \text{ people}) \times \left(3.9 \dfrac{\text{lbm}}{\text{day-person}}\right)\left(365 \dfrac{\text{days}}{\text{yr}}\right)}{1100 \dfrac{\text{lbm}}{\text{yd}^3}} = 115{,}071 \text{ yd}^3/\text{yr}$$

The maximum soil-cover-to-waste ratio, one part soil to $x$ parts waste, is

$$\frac{3{,}170{,}000 \text{ yd}^3}{115{,}071 \dfrac{\text{yd}^3}{\text{yr}} + \dfrac{115{,}071 \dfrac{\text{yd}^3}{\text{yr}}}{x}} = 25 \text{ yr}$$

$$x = 9.81 \quad (9.8)$$

The ratio is 1:9.8.

**The answer is (D).**

**132.** From *NCEES Handbook* Specific Gravity for a Solids Slurry, the weight of dry solids is

$$V = \frac{W_s}{[(100-p)/100]\gamma S}$$

$$W = V\left(\frac{S}{100}\right)\gamma S$$

$$= (1 \text{ m}^3)\left(\frac{34}{100}\right)\left(1000 \dfrac{\text{kg}}{\text{m}^3}\right)(1.16)$$

$$= 406 \text{ kg} \quad (400 \text{ kg})$$

**The answer is (C).**

**133.** Per *NCEES Handbook* Bioconcentration Factor BCF, bioconcentration of toxic contaminants in fish is evaluated using the bioconcentration factor (BCF).

**The answer is (C).**

**134.** From *NCEES Handbook* Time-Weighted Average (TWA),

$$\text{TWA} = \frac{\sum c_i t_i}{\sum t_i}$$

$$\text{TWA}_c = \frac{c_1 t_1 + c_2 t_2 + c_3 t_3}{t_1 + t_2 + t_3}$$

$$\text{TWA}_{\text{acetone}} = \frac{(1080 \text{ ppm})(1 \text{ h}) + (630 \text{ ppm}) \times (2 \text{ h}) + (470 \text{ ppm})(5 \text{ h})}{1 \text{ h} + 2 \text{ h} + 5 \text{ h}}$$

$$= 586 \text{ ppm}$$

$$\text{TWA}_{\text{sec-butanol}} = \frac{(180 \text{ ppm})(1 \text{ h}) + (85 \text{ ppm}) \times (2 \text{ h}) + (15 \text{ ppm})(5 \text{ h})}{1 \text{ h} + 2 \text{ h} + 5 \text{ h}}$$

$$= 53 \text{ ppm}$$

cumulative TWA exposure $= \dfrac{\text{TWA}_1}{\text{TWA-PEL}_1} + \dfrac{\text{TWA}_2}{\text{TWA-PEL}_2}$

$= \dfrac{586 \text{ ppm}}{1000 \text{ ppm}} + \dfrac{53 \text{ ppm}}{150 \text{ ppm}}$

$= 0.94$

*The answer is (B).*

**135.** From *NCEES Handbook* Overall Fan Efficiency, the overall efficiency of the system is

$$\eta_{\text{overall}} = \eta_f \times \eta_t \times \eta_m \times \eta_c$$
$$= (0.93)(0.86)(0.89)(0.95)$$
$$= 0.676 \quad (0.68)$$

*The answer is (A).*

**136.** From *NCEES Handbook* Inverse Square Law, the intensity at 10 m is

$$\dfrac{I_1}{I_2} = \dfrac{(R_2)^2}{(R_1)^2}$$

$$I_1 = \dfrac{I_2 (R_2)^2}{(R_1)^2}$$

$$= \dfrac{\left(0.016 \, \dfrac{\text{R}}{\text{min}}\right)(2 \text{ m})^2 \left(1000 \, \dfrac{\text{mR}}{\text{R}}\right)}{(10 \text{ m})^2}$$

$$= 0.64 \text{ mR/min}$$

*The answer is (B).*

**137.** Chemical A has a low MCL and a low Henry's constant compared to chemicals B, C, and D. The lower MCL requires a higher removal efficiency. The lower Henry's constant makes the removal efficiency more difficult to achieve. Chemical A should be the target contaminant.

*The answer is (A).*

**138.** From the table "Important Families of Organic Compounds" in the *NCEES Handbook*, trichloroethane is a haloalkane, which is a halogenated organic chemical.

From the table "Common Names and Molecular Formulas of Some Industrial (Inorganic and Organic) Chemicals" in the *NCEES Handbook*, the common name for sodium hydroxide is caustic soda, which is a caustic chemical.

From the figure "Hazardous Waste Compatibility Chart" in the *NCEES Handbook*, when mixed, halogenated organics (row 17) and caustics (row 10) react to produce heat (H) and flammable gas (GF).

*The answer is (B).*

**139.** See *NCEES Handbook* Hazardous Waste Characteristics. The waste is hazardous because of corrosivity (pH less than or equal to 2.0) according to 40 CFR 261.22, and does not meet any other classification criteria. If the waste was neutralized to pH greater than 2.0, it would no longer exhibit a characteristic of hazardous waste.

*The answer is (D).*

**140.** From *NCEES Handbook* Inverse Square Law, taking the known distance as the distance from the source at location 1, the distance at which an intensity of 100 mR/h will be measured is

$$\dfrac{I_1}{I_2} = \dfrac{(R_2)^2}{(R_1)^2}$$

$$R_2 = \sqrt{\dfrac{I_1 R_1^2}{I_2}}$$

$$= \sqrt{\dfrac{\left(162{,}000 \, \dfrac{\text{mR}}{\text{hr}}\right)(1.8 \text{ ft})^2}{100 \, \dfrac{\text{mR}}{\text{hr}}}}$$

$$= 72 \text{ ft}$$

*The answer is (D).*

**141.** From *NCEES Handbook* Theim Equation, the equation for the flow rate is

$$Q = \dfrac{2\pi T(h_2 - h_1)}{\ln\left(\dfrac{r_2}{r_1}\right)}$$

The equation for the transmissivity is $T = KD$. Combining the two equations and rearranging, the hydraulic conductivity of the aquifer is

$$Q = \frac{2\pi T(h_2 - h_1)}{\ln\left(\dfrac{r_2}{r_1}\right)}$$

$$= \frac{2\pi KD(h_2 - h_1)}{\ln\left(\dfrac{r_2}{r_1}\right)}$$

$$K = \frac{Q \ln\left(\dfrac{r_2}{r_1}\right)}{2\pi D(h_2 - h_1)}$$

$$= \frac{\left(41 \dfrac{\text{gal}}{\text{min}}\right)\left(0.134 \dfrac{\text{ft}^3}{\text{gal}}\right)\left(1440 \dfrac{\text{min}}{\text{day}}\right)\ln\left(\dfrac{14.8 \text{ ft}}{7.3 \text{ ft}}\right)}{(2)\pi(8.2 \text{ ft})(9.1 \text{ ft} - 5.7 \text{ ft})}$$

$$= 31.9 \text{ ft/day} \quad (32 \text{ ft/day})$$

**The answer is (C).**

**142.** From the table "Water Solubility, Vapor Pressure, Henry's Law Constant, $K_{oc}$, and $K_{ow}$ Data for Selected Chemicals" in the *NCEES Handbook*, the coefficient $K_{oc}$ for ethylbenzene is $1.10 \times 10^3$ mL/g.

From *NCEES Handbook* Soil-Water Partition Coefficient $K_d = K_p$, the soil-water partition coefficient for ethylbenzene in the aquifer is

$$K_d = K_{oc} f_{oc}$$

$$= \left(1.10 \times 10^3 \dfrac{\text{mL}}{\text{g}}\right)(0.32)$$

$$= 352 \text{ mL/g} \quad (350 \text{ mL/g})$$

**The answer is (B).**

**143.** From *NCEES Handbook* Overall Coefficients,

$$C_A^* = p_{AG}/H$$

$$H = \frac{p_{AG}}{C_A^*}$$

$$= \frac{0.26 \text{ atm}}{9300 \dfrac{\text{mg}}{\text{L}}}$$

$$= 2.8 \times 10^{-5} \text{ atm·L/mg}$$

To convert to the unitless form, use the value of the gas constant from *NCEES Handbook* table "Ideal Gas Constants."

$$R = 8.2 \times 10^{-5} \dfrac{\text{atm·m}^3}{\text{mol·K}}$$

$$H_{\text{unitless}} = \frac{\left(2.8 \times 10^{-5} \dfrac{\text{atm·L}}{\text{mg}}\right)\left(120 \dfrac{\text{g}}{\text{mol}}\right)\left(1000 \dfrac{\text{mg}}{\text{g}}\right)}{\left(8.2 \times 10^{-5} \dfrac{\text{atm·m}^3}{\text{mol·K}}\right)(25°\text{C} + 273°)\left(1000 \dfrac{\text{L}}{\text{m}^3}\right)}$$

$$= 0.1375 \quad (0.14)$$

**The answer is (B).**

**144.**

$$\text{EBCT} = \frac{(20 \text{ m}^3)\left(1440 \dfrac{\text{min}}{\text{d}}\right)}{500 \dfrac{\text{m}^3}{\text{d}}}$$

$$= 57.6 \text{ min} \quad (58 \text{ min})$$

**The answer is (B).**

**145.** From *NCEES Handbook* Hydraulic Conductivity, the hydraulic conductivity with respect to the fuel is

$$K = \rho g k / \mu$$

According to *NCEES Handbook* Stress, Pressure, and Viscosity, $\nu = \mu/\rho$, therefore, the equation simplifies to

$$K = \frac{kg}{\nu}$$

Assume kinematic viscosity given is for 8°C.

$$K = \frac{(1.0 \times 10^{-10} \text{ m}^2)\left(9.81 \dfrac{\text{m}}{\text{s}^2}\right)\left(86\,400 \dfrac{\text{s}}{\text{d}}\right)}{\left(8.32 \dfrac{\text{mm}^2}{\text{s}}\right)\left(\dfrac{1 \text{ m}}{1000 \text{ m}}\right)^2}$$

$$= 10.2 \text{ m/d} \quad (10 \text{ m/d})$$

**The answer is (D).**

**146.** See the table "Important Families of Organic Chemicals" in the *NCEES Handbook*. The original compound is 3-pentanol, which belongs to the alcohol family. The compound in the equation formed by the oxidation of 3-pentanol is methyl iso-butanone, also commonly known as methyl isobutyl ketone (MIBK). MIBK belongs to the ketone family.

**The answer is (B).**

**147.** Carbon monoxide will cause death in humans within a few minutes at exposures exceeding 5000 ppm and impair vision, motor function, and other physical and mental abilities at much lower concentrations. These effects occur through formation of carboxyhemoglobin as a reaction product between carbon monoxide and hemoglobin. The carboxyhemoglobin formation effectively acts to deprive the body of oxygen. Heightened nervousness and mental agitation are typically not associated with carbon monoxide exposure.

*The answer is (B).*

**148.** Hazardous air pollutants are those pollutants that are known or suspected to cause cancer or other serious health effects.

*The answer is (C).*

**149.** From *NCEES Handbook* Safety Data Sheet, the SDS would address spill response measures and disposal of wastes generated from spill response, as well as proper storage procedures for the material. Decontamination of wastes generated from spill response do not directly affect workplace safety and would not be included on the SDS.

*The answer is (C).*

**150.** The EPA has regulatory authority for radon gas exposure under the Indoor Radon Abatement Act of 1988.

*The answer is (C).*

**151.** Hydrogen sulfide gas ($H_2S$) produced by the biological reduction of sulfates is the source of one of the most readily recognizable malodors associated with wastewater treatment and is detectable at very low concentrations.

*The answer is (A).*

**152.** From *NCEES Handbook* GHS: Acute Oral Toxicity, the five categories used by GHS for acute oral toxicity are

| category | 1 | 2 | 3 | 4 | 5 |
|---|---|---|---|---|---|
| $LD_{50}$ (mg/kg) | $\leq 5$ | $> 5 < 50$ | $\geq 50 < 300$ | $\geq 300 < 2000$ | $\geq 2000 < 5000$ |

*The answer is (A).*

**153.** From *NCEES Handbook* Half-Life (Radioactive Decay), the half-life is

$$\tau = \frac{0.693}{\lambda}$$

$$\lambda = \frac{0.693}{\tau_{1/2}}$$

$$= \frac{0.693}{28.8 \text{ yr}}$$

$$= 0.024 \text{ yr}^{-1}$$

*The answer is (A).*

**154.** From *NCEES Handbook* Noise Pollution: TWA Noise Level, the time-weighted average (TWA) noise level for a noise dose of 112% is

$$\text{TWA} = 90 + 16.61 \log \frac{D\%}{100\%}$$

$$= 90 \text{ dB} + 16.61 \log \frac{112\%}{100\%}$$

$$= 90.8 \text{ dB} \quad (91 \text{ dB})$$

*The answer is (D).*

**155.**

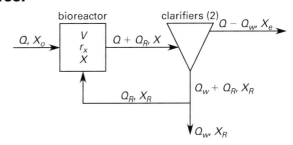

| | | |
|---|---|---|
| $Q$ | influent flow from primary clarifiers | 60 000 m³/d |
| $Q_R$ | return solids flow rate | m³/d |
| $Q_w$ | wasted solids flow rate | m³/d |
| $r_x$ | biomass production rate | 0.5 kg/m³·d |
| $V$ | bioreactor volume | 8000 m³ |
| $X$ | mixed liquor volatile suspended solids (MLVSS) | 1800 mg/L |
| $X_e$ | effluent VSS from secondary clarifiers | mg/L |
| $X_o$ | influent VSS from primary clarifiers | 100 mg/L |
| $X_R$ | return flow VSS | 12 000 mg/L |

Perform a solids mass balance around the bioreactor. Write the statement in words.

influent VSS + return VSS + cell growth
= effluent VSS

Write the equation with variables.

$$QX_o + Q_R X_R + V r_x = (Q + Q_R)X$$

Write the equation with values.

$$\left(60\,000 \ \frac{m^3}{d}\right)\left(100 \ \frac{mg}{L}\right)\left(10^{-3} \ \frac{kg \cdot L}{mg \cdot m^3}\right)$$
$$+ Q_R \left(12\,000 \ \frac{mg}{L}\right)\left(10^{-3} \ \frac{kg \cdot L}{mg \cdot m^3}\right)$$
$$+ (8000 \ m^3)\left(0.5 \ \frac{kg}{m^3 \cdot d}\right)$$
$$= \left(60\,000 \ \frac{m^3}{d} + Q_R\right)\left(1800 \ \frac{mg}{L}\right)$$
$$\times \left(10^{-3} \ \frac{kg \cdot L}{mg \cdot m^3}\right)$$

Multiply through and combine terms.

$$6000 \ \frac{kg}{d} + \left(12 \ \frac{kg}{m^3}\right)Q_R + 4000 \ \frac{kg}{d}$$
$$= 108\,000 \ \frac{kg}{d} + \left(1.8 \ \frac{kg}{m^3}\right)Q_R$$

Rearrange and solve for $Q_R$.

$$108\,000 \ \frac{kg}{d} - 6000 \ \frac{kg}{d} - 4000 \ \frac{kg}{d}$$
$$= \left(12 \ \frac{kg}{m^3} - 1.8 \ \frac{kg}{m^3}\right)Q_R$$

$$Q_R = \frac{98\,000 \ \frac{kg}{d}}{10.2 \ \frac{kg}{m^3}}$$
$$= 9608 \ m^3/d \quad (9600 \ m^3/d)$$

**The answer is (C).**

**156.** The mass of rubber supplied weekly is

$$\left(11\,000 \ \frac{kg}{d}\right)\left(5 \ \frac{d}{wk}\right) = 55\,000 \ kg/wk$$

The mass of rubber supplied hourly to the incinerator is

$$\frac{55\,000 \ \frac{kg}{wk}}{\left(7 \ \frac{d}{wk}\right)\left(24 \ \frac{h}{d}\right)} = 327 \ kg/h$$

The hourly heat value of the shredded rubber is

$$\left(327 \ \frac{kg}{h}\right)\left(90\,000 \ \frac{kJ}{kg}\right) = 2.9 \times 10^7 \ kJ/h$$

**The answer is (B).**

**157.** The streamflow values are ranked from high to low; therefore, the table gives the probability that the flow will be exceeded in any given year.

The flow that has a 20% risk of being equaled or exceeded at least once in 10 years is given by the probability function shown. $P$ is the probability, and $n$ is the duration during which the flow is measured. $P = 1 -$ risk. From *NCEES Handbook* Storm Return Period,

$$P(x) = 1 - P^n = 1 - (1 - 0.2)^{10} = 0.893$$

From the table, the flow value corresponding to 0.893 is approximately 1180 ft$^3$/sec.

**The answer is (A).**

**158.** On day 13, 40% of the work has been completed, 43% of the budget has been expended, and 54% of the schedule has been used. The project is over budget (43% > 40%) and behind schedule (54% > 40%).

**The answer is (D).**

**159.** See *NCEES Handbook* Economics. The gradient is not uniform. The present worth, $P$, is $1,300,000; the duration of the project, $n$, is 3; and the interest rate, $i$, is 2% for capital and 4% for maintenance.

For capital costs, the annual amount is found from the cash flow factor $P(A/P, i, n)$. From interest tables, $(A/P, 2\%, 3) = 0.3468$. The annual amount for capital costs is

$$A = (\$1,300,000)(0.3468) = \$450,840$$

For operations and maintenance (O&M) costs, using interest tables, the present worth in each year of the project is

$245,000(P/F, 4\%, 1) = (\$245,000)(0.9615) = \$235,568$
$209,000(P/F, 4\%, 2) = (\$209,000)(0.9246) = \$193,241$
$167,000(P/F, 4\%, 3) = (\$167,000)(0.8890) = \$148,463$

The total O&M present worth is

$$P = \$235{,}568 + \$193{,}241 + \$148{,}463 = \$577{,}272$$

For the project, the equivalent uniform annual cost is found from the cash flow factor $P(A/P, i, n)$.

From interest tables, $(A/P, 4\%, 3) = 0.3603$. The equivalent uniform annual cost is

$$\begin{aligned} A &= \$577{,}272(A/P, 4\%, 3) + \$450{,}840 \\ &= (\$577{,}272)(0.3603) + \$450{,}840 \\ &= \$658{,}831 \quad (\$660{,}000) \end{aligned}$$

**The answer is (C).**

**160.** The box-and-whisker plot represents the data with the 24-hour mean, mode, inner-quartile range, and extreme values. The bimodality of the traffic data would likely sit in the body of the box and not be apparent.

While a pie chart would show the different traffic volumes measured over the 24-hour cycle, it gives no indication of how one hour is related to the next. A pie chart would not indicate whether traffic is increasing monotonically or otherwise.

Histograms show the frequency/density of the data based on the $x$ variable. For example, a histogram of traffic data would have time, from hour 0 (midnight) until hour 23 (11 PM), and would have a $y$-axis for traffic volume (e.g., the number of cars on the road). A histogram will show spikes in the morning and evening when there are many cars on the road, which will show the bimodality of the data set.

A stacked column chart is very similar to a pie chart, and would not clearly show the two modes in the data set because it does not preserve the relative positions of the data points as a function of time.

**The answer is (C).**